INTRODUCTION TO COMPUTATIONAL CONTACT MECHANICS

WILEY SERIES IN COMPUTATIONAL MECHANICS

Series Advisors:

René de Borst
Perumal Nithiarasu
Tayfun E. Tezduyar
Genki Yagawa
Tarek Zohdi

Introduction to Computational Contact Mechanics: A Geometrical Approach	Konyukhov	April 2015
Extended Finite Element Method: Theory and Applications	Khoei	December 2014
Computational Fluid-Structure Interaction: Methods and Applications	Bazilevs, Takizawa and Tezduyar	January 2013
Introduction to Finite Strain Theory for Continuum Elasto-Plasticity	Hashiguchi and Yamakawa	November 2012
Nonlinear Finite Element Analysis of Solids and Structures, Second Edition	De Borst, Crisfield, Remmers and Verhoosel	August 2012
An Introduction to Mathematical Modeling: A Course in Mechanics	Oden	November 2011
Computational Mechanics of Discontinua	Munjiza, Knight and Rougier	November 2011
Introduction to Finite Element Analysis: Formulation, Verification and Validation	Szabó and Babuška	March 2011

INTRODUCTION TO COMPUTATIONAL CONTACT MECHANICS
A GEOMETRICAL APPROACH

Alexander Konyukhov
Karlsruhe Institute of Technology (KIT), Germany

Ridvan Izi
Karlsruhe Institute of Technology (KIT), Germany

This edition first published 2015
© 2015 John Wiley & Sons Ltd

Registered office
John Wiley & Sons Ltd, The Atrium, Southern Gate, Chichester, West Sussex, PO19 8SQ, United Kingdom

For details of our global editorial offices, for customer services and for information about how to apply for permission to reuse the copyright material in this book please see our website at www.wiley.com.

The right of the author to be identified as the author of this work has been asserted in accordance with the Copyright, Designs and Patents Act 1988.

All rights reserved. No part of this publication may be reproduced, stored in a retrieval system, or transmitted, in any form or by any means, electronic, mechanical, photocopying, recording or otherwise, except as permitted by the UK Copyright, Designs and Patents Act 1988, without the prior permission of the publisher.

Wiley also publishes its books in a variety of electronic formats. Some content that appears in print may not be available in electronic books.

Designations used by companies to distinguish their products are often claimed as trademarks. All brand names and product names used in this book are trade names, service marks, trademarks or registered trademarks of their respective owners. The publisher is not associated with any product or vendor mentioned in this book.

Limit of Liability/Disclaimer of Warranty: While the publisher and author have used their best efforts in preparing this book, they make no representations or warranties with respect to the accuracy or completeness of the contents of this book and specifically disclaim any implied warranties of merchantability or fitness for a particular purpose. It is sold on the understanding that the publisher is not engaged in rendering professional services and neither the publisher nor the author shall be liable for damages arising herefrom. If professional advice or other expert assistance is required, the services of a competent professional should be sought

Library of Congress Cataloging-in-Publication Data

Konyukhov, Alexander.
 Introduction to computational contact mechanics : a geometrical approach / Alexander Konyukhov, Karlsruhe Institute of Technology (KIT), Germany, Ridvan Izi, Karlsruhe Institute of Technology (KIT), Germany.
   pages cm. – (Wiley series in computational mechanics)
 Includes bibliographical references and index.
 ISBN 978-1-118-77065-8 (cloth : alk. paper) 1. Contact mechanics. 2. Mechanics, Applied. I. Izi, Ridvan. II. Title. III. Title: Computational contact mechanics.
 TA353.K66 2015
 620.1′05–dc23

2015005384

A catalogue record for this book is available from the British Library.

Typeset in 11/13pt TimesLTStd by Laserwords Private Limited, Chennai, India

1 2015

Contents

Series Preface   xiii

Preface   xv

Acknowledgments   xix

Part I   THEORY

1   Introduction with a Spring-Mass Frictionless Contact System   3
1.1   Structural Part – Deflection of Spring-Mass System   3
1.2   Contact Part – Non-Penetration into Rigid Plane   4
1.3   Contact Formulations   5
     1.3.1   Lagrange Multiplier Method   5
     1.3.2   Penalty Method   6
     1.3.3   Augmented Lagrangian Method   8

2   General Formulation of a Contact Problem   13
2.1   Structural Part – Formulation of a Problem in Linear Elasticity   13
     2.1.1   Strong Formulation of Equilibrium   14
     2.1.2   Weak Formulation of Equilibrium   15
2.2   Formulation of the Contact Part (Signorini's problem)   17

3   Differential Geometry   23
3.1   Curve and its Properties   23
     3.1.1   Example: Circle and its Properties   26
3.2   Frenet Formulas in 2D   28
3.3   Description of Surfaces by Gauss Coordinates   29
     3.3.1   Tangent and Normal Vectors: Surface Coordinate System   29
     3.3.2   Basis Vectors: Metric Tensor and its Applications   30
     3.3.3   Relationships between Co- and Contravariant Basis Vectors   33
     3.3.4   Co- and Contravariant Representation of a Vector on a Surface   34
     3.3.5   Curvature Tensor and Structure of the Surface   35

3.4	Differential Properties of Surfaces		37
	3.4.1 *The Weingarten Formula*		37
	3.4.2 *The Gauss–Codazzi Formula*		38
	3.4.3 *Covariant Derivatives on the Surface*		38
	3.4.4 *Example: Geometrical Analysis of a Cylindrical Surface*		39

4 Geometry and Kinematics for an Arbitrary Two Body Contact Problem — **45**

4.1 Local Coordinate System — 46
4.2 Closest Point Projection (CPP) Procedure – Analysis — 48
  4.2.1 *Existence and Uniqueness of CPP Procedure* — 49
  4.2.2 *Numerical Solution of CPP Procedure in 2D* — 54
  4.2.3 *Numerical Solution of CPP Procedure in 3D* — 54
4.3 Contact Kinematics — 55
  4.3.1 *2D Contact Kinematics using Natural Coordinates s and ζ* — 58
  4.3.2 *Contact Kinematics in 3D Coordinate System* — 59

5 Abstract Form of Formulations in Computational Mechanics — **61**

5.1 Operator Necessary for the Abstract Formulation — 61
  5.1.1 *Examples of Operators in Mechanics* — 61
  5.1.2 *Examples of Various Problems* — 62
5.2 Abstract Form of the Iterative Method — 63
5.3 Fixed Point Theorem (Banach) — 64
5.4 Newton Iterative Solution Method — 65
  5.4.1 *Geometrical Interpretation of the Newton Iterative Method* — 66
5.5 Abstract Form for Contact Formulations — 69
  5.5.1 *Lagrange Multiplier Method in Operator Form* — 69
  5.5.2 *Penalty Method in Operator Form* — 71

6 Weak Formulation and Consistent Linearization — **73**

6.1 Weak Formulation in the Local Coordinate System — 73
6.2 Regularization with Penalty Method — 75
6.3 Consistent Linearization — 75
  6.3.1 *Linearization of Normal Part* — 76
6.4 Application to Lagrange Multipliers and to Following Forces — 79
  6.4.1 *Linearization for the Lagrange Multipliers Method* — 80
  6.4.2 *Linearization for Following Forces: Normal Force or Pressure* — 80
6.5 Linearization of the Convective Variation $\delta\xi$ — 81
6.6 Nitsche Method — 81
  6.6.1 *Example: Independence of the Stabilization Parameter* — 83

7	**Finite Element Discretization**	**85**
7.1	Computation of the Contact Integral for Various Contact Approaches	86
	7.1.1 Numerical Integration for the Node-To-Node (NTN)	86
	7.1.2 Numerical Integration for the Node-To-Segment (NTS)	86
	7.1.3 Numerical Integration for the Segment-To-Analytical Segment (STAS)	86
	7.1.4 Numerical Integration for the Segment-To-Segment (STS)	87
7.2	Node-To-Node (NTN) Contact Element	88
7.3	Nitsche Node-To-Node (NTN) Contact Element	89
7.4	Node-To-Segment (NTS) Contact Element	91
	7.4.1 Closest Point Projection Procedure for the Linear NTS Contact Element	94
	7.4.2 Peculiarities in Computation of the Contact Integral	95
	7.4.3 Residual and Tangent Matrix	96
7.5	Segment-To-Analytical-Surface (STAS) Approach	98
	7.5.1 General Structure of CPP Procedure for STAS Contact Element	98
	7.5.2 Closed form Solutions for Penetration in 2D	100
	7.5.3 Discretization for STAS Contact Approach	102
	7.5.4 Residual and Tangent Matrix	102
7.6	Segment-To-Segment (STS) Mortar Approach	104
	7.6.1 Peculiarities of the CPP Procedure for the STS Contact Approach	106
	7.6.2 Computation of the Residual and Tangent Matrix	106
8	**Verification with Analytical Solutions**	**109**
8.1	Hertz Problem	109
	8.1.1 Contact Geometry	110
	8.1.2 Contact Pressure and Displacement for Spheres: 3D Hertz Solution	113
	8.1.3 Contact Pressure and Displacement for Cylinders: 2D Hertz Solution	114
8.2	Rigid Flat Punch Problem	114
8.3	Impact on Moving Pendulum: Center of Percussion	116
8.4	Generalized Euler–Eytelwein Problem	118
	8.4.1 A Rope on a Circle and a Rope on an Ellipse	119
9	**Frictional Contact Problems**	**121**
9.1	Measures of Contact Interactions – Sticking and Sliding Case: Friction Law	121
	9.1.1 Coulomb Friction Law	123

9.2 Regularization of Tangential Force and Return Mapping Algorithm 123
    9.2.1 Elasto-Plastic Analogy: Principle of Maximum of Dissipation 125
    9.2.2 Update of Sliding Displacements in the Case of Reversible Loading 127
9.3 Weak Form and its Consistent Linearization 128
9.4 Frictional Node-To-Node (NTN) Contact Element 129
    9.4.1 Regularization of the Contact Conditions 130
    9.4.2 Linearization the of Tangential Part for the NTN Contact Approach 131
    9.4.3 Discretization of Frictional NTN 131
    9.4.4 Algorithm for a Local Level Frictional NTN Contact Element 133
9.5 Frictional Node-To-Segment (NTS) Contact Element 134
    9.5.1 Linearization and Discretization for the NTS Frictional Contact Element 134
    9.5.2 Algorithm for a Local Level NTS Frictional Contact Element 135
9.6 NTS Frictional Contact Element 135

Part II PROGRAMMING AND VERIFICATION TASKS

10 Introduction to Programming and Verification Tasks **139**

11 Lesson 1 Nonlinear Structural Truss – elmt1.f **143**
11.1 Implementation 144
11.2 Examples 148
    11.2.1 Constitutive Laws of Material 148
    11.2.2 Large Rotation 149
    11.2.3 Snap-Through Buckling 150

12 Lesson 2 Nonlinear Structural Plane – elmt2.f **151**
12.1 Implementation 152
12.2 Examples 156
    12.2.1 Constitutive Law of Material 156
    12.2.2 Large Rotation 158

13 Lesson 3 Penalty Node-To-Node (NTN) – elmt100.f **159**
13.1 Implementation 160
13.2 Examples 161
    13.2.1 Two Trusses 161
    13.2.2 Three Trusses 162
    13.2.3 Two Blocks 163

14 Lesson 4 Lagrange Multiplier Node-To-Node (NTN) – elmt101.f **165**
14.1 Implementation 166

14.2	Examples	168
	14.2.1 Two Trusses	168
	14.2.2 Three Trusses	169

15 Lesson 5 Nitsche Node-To-Node (NTN) – `elmt102.f` — 171
15.1	Implementation	171
15.2	Examples	174
	15.2.1 Two Trusses	174
	15.2.2 Three Trusses	174

16 Lesson 6 Node-To-Segment (NTS) – `elmt103.f` — 177
16.1	Implementation	178
16.2	Examples	181
	16.2.1 Two Blocks	181
	16.2.2 Two Blocks – Horizontal Position	182
	16.2.3 Two Cantilever Beams – Large Sliding Test	183
	16.2.4 Hertz Problem	183
16.3	Inverted Contact Algorithm – Following Force	185
	16.3.1 Verification of the Rotational Part – A Single Following Force	186

17 Lesson 7 Segment-To-Analytical-Segment (STAS) – `elmt104.f` — 189
17.1	Implementation	190
17.2	Examples	193
	17.2.1 Block and Rigid Surface	193
	17.2.2 Block and Inclined Rigid Surface	194
	17.2.3 Block and Inclined Rigid Surface – different Boundary Condition	195
	17.2.4 Bending Over a Rigid Cylinder	196
17.3	Inverted Contact Algorithm – General Case of Following Forces	196
	17.3.1 Verification of a Rotational Part – A Single Following Force	198
	17.3.2 Distributed Following Forces – Pressure	199
	17.3.3 Inflating of a Bar	201

18 Lesson 8 Mortar/Segment-To-Segment (STS) – `elmt105.f` — 203
18.1	Implementation	204
18.2	Examples	207
	18.2.1 Two Blocks	207
	18.2.2 Block and Inclined Rigid Surface – Different Boundary Condition	208
	18.2.3 Contact Patch Test	209
18.3	Inverted Contact Algorithm – Following Force	210
	18.3.1 Verification of the Rotational Part – Pressure on the Master Side	211

19	**Lesson 9 Higher Order Mortar/STS – elmt106.f**	**213**
19.1	Implementation	214
19.2	Examples	217
	19.2.1 Two Blocks	218
	19.2.2 Block and Inclined Rigid Surface – Different Boundary Condition	219
20	**Lesson 10 3D Node-To-Segment (NTS) – elmt107.f**	**221**
20.1	Implementation	222
20.2	Examples	225
	20.2.1 Two Blocks – 3D Case	226
	20.2.2 Sliding on a Ramp	226
	20.2.3 Bending Over a Rigid Cylinder	227
	20.2.4 Bending Over a Rigid Sphere	227
21	**Lesson 11 Frictional Node-To-Node (NTN) – elmt108.f**	**229**
21.1	Implementation	230
21.2	Examples	232
	21.2.1 Two Blocks – Frictional Case	232
	21.2.2 Frictional Contact Patch Test	233
22	**Lesson 12 Frictional Node-To-Segment (NTS) – elmt109.f**	**235**
22.1	Implementation	236
22.2	Examples	240
	22.2.1 Two Blocks	240
	22.2.2 Frictional Contact Patch Test	241
	22.2.3 Block and Inclined Rigid Surface – Different Boundary Condition	242
	22.2.4 Generalized 2D Euler–Eytelwein Problem	243
23	**Lesson 13 Frictional Higher Order NTS – elmt110.f**	**245**
23.1	Implementation	246
23.2	Examples	250
	23.2.1 Two Blocks	251
	23.2.2 Block and Inclined Rigid Surface – Different Boundary Condition	252
24	**Lesson 14 Transient Contact Problems**	**255**
24.1	Implementation	256
24.2	Examples	257
	24.2.1 Block and Inclined Rigid Surface – Non-Frictional Case	257
	24.2.2 Block and Inclined Rigid Surface – Frictional Case	258
	24.2.3 Moving Pendulum with Impact – Center of Percussion	258

Appendix A Numerical integration — **261**
A.1 Gauss Quadrature — 262
    A.1.1 Evaluation of Integration Points — 262
    A.1.2 Numerical Examples — 263

Appendix B Higher Order Shape Functions of Different Classes — **265**
B.1 General — 265
B.2 Lobatto Class — 265
    B.2.1 1D Lobatto — 265
    B.2.2 2D Lobatto — 266
    B.2.3 Nodal FEM Input — 269
B.3 Bezier Class — 269
    B.3.1 1D Bezier — 269
    B.3.2 2D Bezier — 270
    B.3.3 Nodal FEM Input — 272

References — **273**

Index — **275**

Series Preface

Since the publication of the seminal paper on contact mechanics by Heinrich Hertz in 1882, the field has grown into an important branch of mechanics, mainly due to the presence of a high number of applications in many branches of engineering. The advent of computational techniques to handle contact between deformable bodies has greatly enhanced the possibility of analyzing contact problems in detail, resulting, for instance, in an enhanced insight into wear problems. The numerical treatment of contact belongs to the hardest problems in computational engineering, and many publications and books have been written to date, marking progress in the field. An *Introduction to Computational Contact Mechanics: A Geometrical Approach* stands out in terms of the clear and geometric approach chosen by the authors. The book covers many aspects of computational contact mechanics and benefits from clear notation. It comes with detailed derivations and explanations, and an exhaustive number of programming and verification tasks, which will help the reader to master the subject.

Preface

Computational contact mechanics within the last decade has developed into a separate branch of computational mechanics dealing exclusively with the numerical modeling of contact problems. Several monographs on computational contact mechanics summarize the study of computational algorithms used in the computational contact mechanics. The most famous, and subject to several editions, are the monographs by Wriggers (2002) and Laursen (2002). Most of the topics are explained at a high research level, which requires a very good knowledge of both numerical mathematics and continuum mechanics. Therefore, this book was the idea of Professor Dr. Ing. Schweizerhof back in 2006, who proposed to me to introduce a course in computational contact mechanics in such a manner that the prerequisite knowledge should be minimized. The main goal was to explain many algorithms used in well-known Finite Element Software packages (ANSYS, ABAQUS, LS-DYNA) in a simple manner and learn their finite element implementation. The starting point of the course was a reduction of the original 3D finite element algorithms into the 2D case and an introductory part to differential geometry. As field of the research has developed, the exploitation of the geometrical methods, the so-called covariant approach, after years of research has lead to the joint research monograph together with Professor Dr. Ing. Schweizerhof in Konyukhov and Schweizerhof (2012) *Computational Contact Mechanics: Geometrically Exact Theory for Arbitrary Shaped Bodies*. At this point, we would like to mention other monographs that we recommend for reading in computational contact mechanics Kikuchi (1988), Sofonea (2012), Yastrebov (2013), in friction and tribology Popov (2010) and also, the famous book on analytical methods in contact mechanics by Johnson (1987).

Ridvan Izi joined the computational contact mechanics course in 2009 and started to give assistance from 2011 to the exercise programming part, and made a lot of effort to make the exercises "easy going" for the students. Thus, the joint work started, leading to the current structure of the exercises in Part II. We were trying to keep this structure independent as much as possible from the programming language, although the course has been given in FEAP (Finite Element Analysis Program) written in FORTRAN.

The current book is based on the course being taught over several years at the Karlsruhe Institute of Technology and proved to be an effective guide for graduate and PhD students studying computational contact mechanics. The geometrically exact theory for contact interaction is delivered in a simple attractive engineering manner available for undergraduate students starting from 1D geometry.

The book is subdivided into two parts:

- Part I contains the theoretical basis for the computational contact mechanics, including necessary material for lectures in computational contact mechanics.
- Part II includes the necessary material for the practical implementation of algorithms, including verification and numerical analysis of contact problems. Part II is consequently constructed following the theory considered in Part I.
- In addition, the original FORTRAN programs, including all numerical examples considered in Part II, are available from the supporting Wiley website at www.wiley.com/go/Konyukhov

The basis of the geometrically exact theory for contact interaction is to build the proper coordinate system to describe the contact interaction in all its geometrical detail. This results in the special structure of the computational mechanics course – study in applied differential geometry, kinematics of contact, formulation of a weak form and linearization in a special coordinate system in a covariant form. Afterward, most popular methods to enforce contact conditions – the penalty method, Lagrange multipliers, augmented Lagrange multipliers, Mortar method and the more seldomly used Nitsche method – are formulated consequently, first for 1D and then for 2D systems finally leading to examples in 3D. It then applies to finite element discretization. The structure of contact elements for these methods is studied in detail and all numerical algorithms are derived in a form ready for implementation. Thus, the structure of contact elements is carefully derived for various situations: Node-To-Node (NTN), Node-To-Segment (NTS) and Segment-To-Segment (STS) contact approaches. Special attention is given to the derivation of contact elements with rigid bodies of simple geometry such as the Segment-To-Analytical Segment (STAS) approach.

Part II of the book contains programming schemes for the following finite elements: surface-to-analytical (rigid) surface, NTN for several methods: penalty, Lagrange, Nitsche methods; node-to-segment for both non-frictional and frictional cases, with Mortar type segment-to-segment and 3D node-to-segment contact elements. Through examples, special attention is given to the implementation of normal following forces, which is derived a particular case of implementation for the frictionless contact algorithm.

All examples are given in a sequential manner with increasing complexity, which allows the reader to program these elements easily. Though the course has been designed for the FEAP user using FORTRAN, the structure of all examples is given in a programming-block manner, which allows the user to program all elements using any, convenient programming language or just mathematical software such as MATLAB.

The examples and corresponding tests are conceptualized in order to study many numerical phenomena appearing in computational contact mechanics, such as influence of the penalty parameters, selection of meshes and element type for the contact patch test for non-frictional and frictional cases.

The original implementation of the derived contact elements was carried out in one of the earliest versions of FEAP originating from Professor Robert Taylor, University of California, Berkley. The Finite Element Analysis Program (FEAP) appeared at the Institute of Mechanics of the Karlsruhe Institute of Technology due to the joint collaboration between Robert Taylor and Karl Schweizerhof who further developed the FEAP code into FEAP-MeKa with the famous solid-shell finite element. The code used in the current course is a simplified student version without any finite elements used for research and is used for educational purposes. During private communications, Professor Taylor confirmed that a free updated version is available and is still supported at http://www.ce.berkeley.edu/projects/feap/feappv/. I am particularly thankful for his kind agreement to link the programming given in the current course to the updated version of FEAP. Though all originally implemented subroutines for contact elements are shown within the old version of FEAP (or FEAP-MeKa) together with all necessary specifications (geometry, loads, boundary conditions, etc.) of tasks, the subroutines can be easily rearranged for the updated version of FEAP. The code, together with numerical examples, is essential in order to work with examples given in Part II. Any reader familiar with FEAP can straightforwardly adopt this code to his/her needs. The code is written in FORTRAN, but the straightforward programming structure, without using any math library, is intentionally preserved in order that any user can easily adopt the code to any other programming language. Moreover, we do really hope that the flowcharts, provided for each contact element can be used for programming of computational contact mechanics exercises using symbolic mathematical software such as MATLAB, MATHEMATICA, and so on.

Alexander Konyukhov
Karlsruhe Institute of Technology
Germany

Acknowledgments

We are thankful to Professor Karl Schweizerhof for giving us the great opportunity to develop such a course for students.

We would like to thank Professor Robert Taylor for his kind agreement to link our course of computational contact mechanics to the current and updated version of FEAP, thus encouraging us to work with the current code. Professor Taylor confirmed that the free version is available and is still supported at http://www.ce.berkeley.edu/projects/feap/feappv/.

The group of excellent student assistants has been busy carefully testing all examples given in Part II in less than a year. We would particularly like to thank Christian Lorenz, Merita Haxhibeti, Isabelle Niesel and Oana Mrenes for careful testing of the contact mechanics examples and Marek Fassin for testing necessary structural finite elements. An additional thanks to Oana Mrenes for the many editing efforts made with contemporary LATEX packages.

Many thanks to Johann Bitzenbauer for the careful reading of the current manuscript version and his fruitful proposals that lead to improvement.

The book in its current version has been tested in a workshop for computational contact mechanics recently at the Bundeswehr Universität München – and we are thankful to Georgios Michaloudis for his careful reading and proposals.

The work on this book took us many weekends, sacrificing time spent with our families. At this last, but not least point, we would like to especially thank our families for their understanding and moral support during the work on this book.

Part I
Theory

Part I

Theory

1

Introduction with a Spring-Mass Frictionless Contact System

We start our introduction to contact mechanics from the simplest possible system: a mass point suspended on a spring, but free deformations of the system can be restricted by the additional plane. This chapter gives the general idea on how to handle contact by specifying contact constraints as well as illustrating numerical methods: Lagrange multipliers, Penalty and Augmented Lagrangian.

Contact from geometrical point of view can be observed as a restriction on certain motions. The most simple system to start, shown in Wriggers (2002), is a spring-mass system. Thus, a mass point m suspended on the spring with stiffness c gives us the simplest example of the contact problem, if we restrict the motion (deflection of the spring u) of the point by the rigid plane below, see Figure 1.1.

1.1 Structural Part – Deflection of Spring-Mass System

The equilibrium state of the given system, first without contact, can be described by the following three equivalent expressions:

1. Strong formulation or just an equation for equilibrium of forces

$$F = mg$$

   with spring force represented by $F = cu$ gives us

$$cu - mg = 0; \qquad (1.1)$$

2. Weak or variational formulation is obtained if we consider the work of forces over small test displacement δu (variation).

$$\delta u (cu - mg) = 0 \qquad (1.2)$$

Introduction to Computational Contact Mechanics: A Geometrical Approach, First Edition.
Alexander Konyukhov and Ridvan Izi.
© 2015 John Wiley & Sons, Ltd. Published 2015 by John Wiley & Sons, Ltd.
Companion Website: www.wiley.com/go/Konyukhov

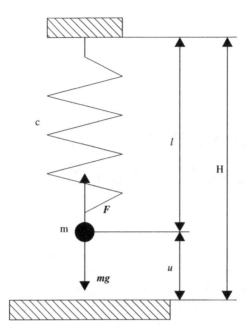

Figure 1.1 Spring-mass system in contact with a rigid plane

δu is also called virtual displacement or the variation of the displacement u. Treating the variation operation δ in a similar way to differentiation, from equation (1.2) we obtain by the integration the following expression

$$\delta \left(\frac{1}{2} c u^2 - m g u \right) = \delta \Pi = 0 \qquad (1.3)$$

with Π representing the potential energy of the system. Equation (1.3) expresses in due course the necessary condition of the extremum and one can see from the positive polynomial of the second order for Π in equation (1.4) that the potential energy Π reaches its minimum. Thus we obtain;

3. Weak or variational formulation in extremal form

$$\Pi(u) = \frac{1}{2} c u^2 - m g u \rightarrow min \qquad (1.4)$$

Namely, the last formulation in the extremal form will be employed to include contact conditions in that form of constraints using methods of optimization theory.

1.2 Contact Part – Non-Penetration into Rigid Plane

The non-penetration condition can be expressed by the geometrical inequality using the total height H of the constrained spring-mass system and the undeformed length of the spring l (undeformed). Thus, expression

$$l + u \leq H$$

Introduction with a Spring-Mass Frictionless Contact System

follows. Equivalently, this can be expressed using a penetration function as $p(u)$

$$p(u) := l + u - H \leq 0. \tag{1.5}$$

Using methods of the optimization theory it is possible to solve the minimization problem $\Pi \rightarrow min$ subjected to such a restriction of p in equation (1.5). Summarizing all diversities of contact situations in the sense of optimization theory, so-called, the Karush–Kuhn–Tucker conditions (or KKT-conditions) are formed:

$$\text{contact:} \quad p = 0 \text{ and } N > 0 \tag{1.6}$$
$$\text{no contact:} \quad p < 0 \text{ and } N = 0$$
$$\text{complimentary condition:} \; pN = 0$$

Formally, N is introduced as a Lagrange multiplier, however, with the mechanical interpretation as the normal contact force. The solution of the contact problem is formulated as minimizing the potential energy in equation (1.4) subject to the inequality conditions (KKT-conditions) in equation (1.6).

1.3 Contact Formulations

Various contact formulations can be obtained depending on the method to satisfy the inequality conditions. The most common methods are the Lagrange multiplier method, the Penalty method and the Augmented Lagrangian method, which are discussed in the following sections for the aforementioned spring-mass contact system.

1.3.1 Lagrange Multiplier Method

The Lagrange multiplier method is based on the construction of a Lagrange functional including constraints. The Lagrange functional L is constructed as a goal function – now potential energy Π – plus constraint equations times the Lagrange multiplier λ

$$L(u, \lambda) = \Pi + \lambda p(u). \tag{1.7}$$

The new functional $L(u, \lambda)$ is, hereby, not dependent on just u any more, but also on λ. Thus, the weak formulation in the form of the extremal problem in equation (1.4) is reformulated as

$$L(u, \lambda) = \Pi + \lambda p \rightarrow min.$$

The necessary condition for fulfilling this requirement is that the partial derivatives with respect to both variables u and λ have to be zero:

$$\begin{cases} \dfrac{\partial L}{\partial u} = 0 \Rightarrow \dfrac{\partial}{\partial u}\left(\dfrac{cu^2}{2} - mgu\right) + \lambda\dfrac{\partial}{\partial u}(l + u - H) = 0 \\ \dfrac{\partial L}{\partial \lambda} = 0 \Rightarrow \qquad\qquad\qquad p = l + u - H = 0. \end{cases}$$

The following system is derived

$$\begin{cases} cu - mg + \lambda = 0 \\ l + u - H = 0. \end{cases} \quad (1.8)$$

The resulting two equations can be interpreted as that the Lagrange multiplier λ physically represents a force that equals the normal contact force ($\lambda = N$). Moreover, the non-penetration condition $p = 0$ is fulfilled exactly. The Karush–Kuhn–Tucker conditions are expressed as follows:

1. contact
   $p = 0$ and $\lambda > 0$;
2. no contact
   $p < 0$ and $\lambda = 0$;
3. complimentary condition
   $p\lambda = 0$
   either ($p = 0$ and $\lambda \neq 0$), or ($p \neq 0$ and $\lambda = 0$).

Remark 1.3.1 *The Lagrange multipliers method allows fulfillment the non-penetration condition exactly, however, the global system needs to be extended by an additional variable – the Lagrange multiplier λ. This can be regarded as an additional obstacle to solving the contact problem.*

1.3.2 Penalty Method

The Penalty method is based on the construction of a new functional without introducing any additional unknown, such as a Lagrange multiplier λ. The functional in equation (1.4) is extended by a penalty functional constructed with a penalty parameter ε and the penetration function $p(u)$

$$\Pi_p(u) = \Pi + \frac{1}{2}\varepsilon p^2(u). \quad (1.9)$$

The additional term $\frac{1}{2}\varepsilon p^2(u)$ is a penalty functional $W_\varepsilon(p)$ depending on both the penetration function p and the penalty parameter ε. We are now going to show that the contact problem will be solved if we increase the penalty parameter to infinity $p \to \infty$.

This is not the only possibility for formulating the penalty functional $W_\varepsilon(p)$ in equation (1.9). It has been proved in numerical methods for optimization that the penalty functional, in general, should fulfill the following conditions:

1. $W_\varepsilon(0) = 0$
2. $W_\varepsilon(p) > 0$ and strictly increases for both p and ε.
3. $\lim_{p \to \infty} W_\varepsilon(p) = \infty$.

Introduction with a Spring-Mass Frictionless Contact System

As an example of the penalty functional, fulfilling these mentioned conditions, it can be constructed in various forms as a positive Taylor series, for example

$$W_\varepsilon(p) = \frac{1}{2}\varepsilon p^2 + \varepsilon_4 p^4 + \ldots \quad (1.10)$$

These conditions are necessary for the fulfillment of constraints in limit case when $p \to 0$ if $\varepsilon(\varepsilon_4,\ldots) \to \infty$.

We have to note that for this contact formulation the functional Π_p depends only on the primary variable u and the choice of the penalty parameter. Thus, the necessary condition for gaining a minimum for Π_p is only one equation as

$$\frac{\partial \Pi_p}{\partial u} = cu - mg + \frac{1}{2}\varepsilon\frac{\partial}{\partial u}(l+u-H)^2 = cu - mg + \varepsilon(l+u-H) = 0 \quad (1.11)$$

Here again, an interpretation for the normal contact force N can be gained as

$$N = \varepsilon(l+u-H) = \varepsilon p(u). \quad (1.12)$$

The last equation gives the mechanical interpretation of the penalty method, see Figure 1.2. In order to fulfill the non-penetration condition an additional spring with a stiffness ε is added at the point of penetration, thus leading to the pulling backwards force $N = \varepsilon p(u)$. It is obvious that the choice of a spring with the infinite

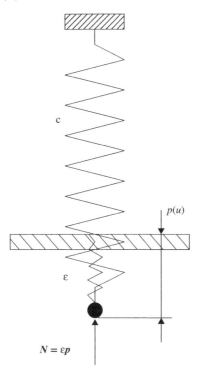

Figure 1.2 Mechanical interpretation of the Penalty method

stiffness leads to zero deformation of the additional spring, or to zero penetration. This backward pulling force is a contact force. This action is comparable to a spring with a stiffness ε and acting on the deflection $p = l + u - H$ attached to the rigid plane. There is no further restriction for the non-penetration condition $p = 0$ and thus penetration $p(u) > 0$ is allowed.

Let us study the deflection u at the limit case $\varepsilon \to \infty$

$$\lim_{\varepsilon \to \infty} u = \lim_{\varepsilon \to \infty} \frac{\varepsilon(H - l) + mg}{c + \varepsilon} = \lim_{\varepsilon \to \infty} \frac{H - l + \frac{mg}{\varepsilon}}{\frac{c}{\varepsilon} + 1} = H - l. \qquad (1.13)$$

The last means exactly that penetration in equation (1.5) vanishes if $\varepsilon \to \infty$

$$\lim_{\varepsilon \to \infty} p(u) = l + u - H = 0.$$

Remark 1.3.2 *The following remarks should be made:*

1. *The non-penetration condition is fulfilled exactly $p = 0$ if the penalty parameter ε goes to infinity.*
2. *In practical computations, "large values" of the penalty parameter ε are considered. This leads to "acceptable" small values for the penetration p. This will be studied during numerical tests in Part II.*
3. *The Penalty method fulfills the non-penetration condition only approximately, however, the global system is not extended and contains only the primary variable u.*

1.3.3 Augmented Lagrangian Method

The augmented Lagrangian method is introduced as an alternative approach to the Lagrange multipliers method and the Penalty method in order to find a way to fulfill the non-penetration condition exactly, however, without introducting an additional variable – Lagrange multiplier λ. The augmented Lagrange method is based on the following functional:

$$L(u) = \Pi + \lambda p(u) + \frac{1}{2}\varepsilon p^2(u). \qquad (1.14)$$

Although the functional $L(u)$ contains terms introduced before in paragraphs 1.3.1 and 1.3.2 for both the Lagrange and the Penalty method with parts $\lambda p(u)$ and $\frac{1}{2}\varepsilon p^2(u)$, respectively, a special recursive scheme is constructed such that the functional depends only on the primary variable u. This fact results in a applied recursive scheme for computing the Lagrange multiplier λ in an iterative (repetitive) manner. This procedure is called augmentation of the Lagrange multiplier, hence the name of the method. The scheme consists of two main steps:

Step 1: Initialization of the Lagrange multiplier λ
$\lambda_0 = 0$
Definition of the initial displacement u_0 by using the system as for the Penalty method equation (1.11)
$cu_0 - mg + \varepsilon(l + u_0 - H) = 0$

Introduction with a Spring-Mass Frictionless Contact System

Step 2: Solution of the minimization problem only via the primary variable u for the Lagrangian in equation (1.14)

$$\frac{\partial L}{\partial u} = \frac{\partial \Pi}{\partial u} + \lambda + \varepsilon p = 0 \qquad (1.15)$$

using the following recursive scheme-augmentation

$$\begin{cases} \lambda_k = \lambda_{k-1} + \varepsilon(l + u_{k-1} - H) \\ cu_k - mg + \lambda_k + \varepsilon(l + u_k - H) = 0, \end{cases} \quad k = 1, 2, 3, \ldots \qquad (1.16)$$

where k is an augmentation counter.

Equation (1.16) will be justified in the following theorem.

Theorem 1.3.3 *The augmented Lagrangian scheme equation (1.16) converges where $\varepsilon > 0$ in the following sense*

$$\lim_{k \to \infty} p(u_k) = 0, \qquad (1.17)$$

Proof. The statement $\lim_{k \to \infty} p(u_k) = 0$ is proved using the analogy between the finite difference equations (FDE) and the ordinary differential equations (ODE). Using the recursive scheme equation (1.16) for iteration step k and $k+1$ of the primary variable u_{k-1} we can write the following sequence

$$\lambda_{k+1} = \lambda_k + \varepsilon(l + u_k - H) \qquad (1.18)$$

$$cu_k - mg + \lambda_k + \varepsilon(l + u_k - H) = 0 \qquad (1.19)$$

$$cu_{k+1} - mg + \lambda_{k+1} + \varepsilon(l + u_{k+1} - H) = 0 \qquad (1.20)$$

Substituting λ_{k+1} and λ_k in equation (1.18) via equations (1.19) and (1.20) we obtain after transformations

$$(c + \varepsilon)(u_{k+1} - u_k) = -\varepsilon(l - H) - \varepsilon u_k. \qquad (1.21)$$

The last is a finite difference equation. Considering the analogy between FDE and ODE by interpreting the iteration step k as the time t, equation (1.21) can be written as

$$(c + \varepsilon)(u_{t+1} - u_t) = -\varepsilon(l - H) - \varepsilon u_t. \qquad (1.22)$$

By dividing the equation then with $\Delta t = 1$ and replacing the difference by $\Delta u = u_{t+1} - u_t$ the following equation is obtained

$$(c + \varepsilon)\frac{\Delta u}{\Delta t} + \varepsilon u_t = -\varepsilon(l - H)$$

where the assumption of the finite difference scheme for the first derivative holds ($\frac{\Delta u}{\Delta t} \approx \frac{du}{dt} = \dot{u}$). Thus, a proof for $k \to \infty$ for FDE can be reformulated in the sense ODE as $t \to \infty$ for the differential equation

$$(c + \varepsilon)\dot{u} + \varepsilon u = -\varepsilon(l - H) \quad \text{with the initial condition} \quad u|_{t=0} = u_0. \quad (1.23)$$

The solution of this non-homogenous ODE is given as

$$u = u_h + u_{nh} \quad (1.24)$$

The solution of the homogenous ODE u_h

$$(c + \varepsilon)\dot{u} + \varepsilon u = 0 \quad (1.25)$$

is obtained according to the characteristic polynomial

$$\lambda(c + \varepsilon) + \varepsilon = 0 \to \lambda = -\frac{\varepsilon}{c + \varepsilon} \quad (1.26)$$

This results in the exponential solution

$$u_h = Ae^{\lambda t} = Ae^{-\frac{\varepsilon}{c+\varepsilon}t}. \quad (1.27)$$

The particular solution of the non-homogeneous ODE u_{nh} is derived as

$$\varepsilon u_{nh} = -\varepsilon(l - H) \quad \Rightarrow \quad u_{nh} = H - l. \quad (1.28)$$

The full general solution is obtained as

$$u = Ae^{-\frac{\varepsilon}{c+\varepsilon}t} + H - l \quad (1.29)$$

which has to fulfill the initial condition $u|_{t=0} = u_0$ defining A as

$$u_0 = A + H - l.$$

Finally, the solution of our problem satisfying the initial condition is derived as

$$u = (u_0 - H + l)e^{-\frac{\varepsilon}{c+\varepsilon}t} + H - l. \quad (1.30)$$

Now we are coming back to the initial proof with the limit process if $t \to \infty$. In this case the counter of augmentations goes to infinity as $k \to \infty$. The limit of equation (1.30) is easily calculated

$$\lim_{t \to \infty} u = H - l \quad \text{for } c, \varepsilon > 0. \quad (1.31)$$

Thus, the penetration function in equation (1.5) is zero in this limit process

$$\lim_{t \to \infty} p(u) = 0 \quad \Rightarrow \quad \lim_{k \to \infty} p(u_k) = 0. \quad (1.32)$$

We prove that the augmented Lagrangian method is independent of the positive penalty parameter $\varepsilon > 0$ and converges to the exact non-penetration condition ($p(u) = 0$) if the number of augmentations goes to infinity $k \to \infty$.

Remark 1.3.4 *In practical computation the iteration is continued until the penetration is smaller than some prescribed tolerance variable ϵ_{tol}*

$$|p_k| = |l + u_k - H| < \epsilon_{tol} \tag{1.33}$$

Remark 1.3.5 *The augmented Lagrangian method fulfills the non-penetration condition within the prescribed tolerance ϵ_{tol}, which is the second parameter of the iterative scheme (first parameter to supply is the penalty parameter ε). The global system is not extended and contains only the primary variable u.*

Remark 1.3.6 *The contact force on each augmentation step k is computed as the augmented Lagrange multiplier in equation (1.18): $N_k = \lambda_k$.*

2

General Formulation of a Contact Problem

The formulation of contact conditions for the simplest spring-mass system can be easily extended into continuum mechanics. The general scheme to formulate the contact problem with the application of further solution methods can be summarized as follows:

1. Formulation of equilibrium equations for contacting bodies – the strong formulation.
2. Formulation of equilibrium condition in a variational, or a weak form. The weak formulation leads to the extremal problem.
3. Formulation of contact conditions in the form of inequalities. Karush–Kuhn–Tucker (KKT) conditions for contact.
4. Formulation of the contact problem as the optimization problem with inequality constraints.
5. Application/selection of a method to solve the constrained optimization problem.

In this chapter, we consider the formulation of the contact problem in continuum mechanics.

2.1 Structural Part – Formulation of a Problem in Linear Elasticity

We outline here the necessary steps from continuum mechanics that are applicable to linear elasticity following the aforementioned scheme.

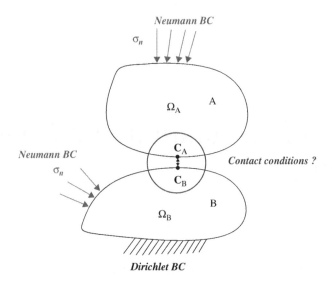

Figure 2.1 Source of nonlinearity of contact problem: neither contact stresses, nor contact surface are known

2.1.1 Strong Formulation of Equilibrium

Equilibrium equation for a 3D continuum domain with Ω representing the body (see Figure 2.1) is formulated for each point $x = (x^1, x^2, x^3), \in \Omega$ as a set of partial differential equations for stress tensor components σ^{ij}

$$\nabla_i \sigma^{ij} = -f^j, \quad i, j = 1, 2, 3. \tag{2.1}$$

Definition 2.1.1 *Here, we introduce the Nabla operator ∇_i in the Cartesian coordinate system, written as a partial derivative with respect to coordinates*

$$\nabla_i(\ldots) \equiv \frac{\partial(\ldots)}{\partial x^i}, i = 1, 2, 3.$$

For an arbitrary curvilinear coordinate system, the Nabla operator will contain the covariant derivatives, see equation (3.67)

Definition 2.1.2 *The Einstein notation, or Einstein summation convention, is used here in order to shorten the writing of the summation: the summation is assumed by repeated indexes – in tensor analysis – one lower and one upper index. In writing equation (2.1) as*

$$\nabla_i \sigma^{ij} \equiv \sum_{i=1}^{3} \nabla_i \sigma^{ij} = \nabla_1 \sigma^{1j} + \nabla_2 \sigma^{2j} + \nabla_3 \sigma^{3j}.$$

Kinematics of deformation describes the relationship between a displacement vector with components u_i and deformations, as well as the chosen measures of

General Formulation of a Contact Problem

deformation. The *linear infinitesimal stain tensor* is used to describe kinematic relationships in the case of geometrically linear deformations:

$$\varepsilon_{ij} = \frac{1}{2}(\nabla_i u_j + \nabla_j u_i). \tag{2.2}$$

The system of partial differential equations in (2.1) describes equilibrium conditions for the continuum independent of the mechanical properties of the material filling this continuum. Thus, *constitutive equations* are necessary then to describe the material model by setting dependency of the stress tensor from the strain tensor $\sigma^{ij} = \sigma^{ij}(\varepsilon_{ij})$. The linear Hook's law is employed to specify the *linear elastic material* in general anisotropic case as

$$\sigma^{ij} = A^{ijkl}\varepsilon_{kl}, \quad i,j = 1,2,3. \tag{2.3}$$

For the formulation of equilibrium conditions in the form of the boundary value problem (BVP), the following boundary conditions can be considered, see Figures 2.1 and 2.2:

1. Neumann boundary conditions – given surface forces $\sigma^i_{(n)}$ on the boundary surface Γ_σ (the Cauchy stress theorem);

$$\sigma^{ij}n_j = \sigma^i_{(n)} \text{ for } x \in \Gamma_\sigma; \tag{2.4}$$

2. Dirichlet boundary conditions – either given prescribed displacements or just fixed boundary of the boundary surface Γ_u

$$u^i = u^i_0 \text{ for } x \in \Gamma_u. \tag{2.5}$$

2.1.2 Weak Formulation of Equilibrium

The weak form of the equilibrium is derived by multiplying equation (2.1) with test functions (virtual displacements or variation of displacements) δu_j, satisfying Dirichlet boundary conditions on the part of the surface Γ_u, summing over $j = 1,2,3$ and then integrating over the domain Ω

$$\int_\Omega \nabla_i \sigma^{ij} \delta u_j d\Omega + \int_\Omega f^j \delta u_j d\Omega = 0. \tag{2.6}$$

The equilibrium condition in the weak formulation has no longer to be fulfilled for all points of Ω, but in an average manner in the domain Ω. The used test functions δu_j have to satisfy the following properties:

1. smooth (derivatives exist);
2. $\delta u_j|_{\Gamma_u} = 0$ (vanish at Dirichlet boundaries).

For the following transformation we recall the Gauss theorem as a lemma here:

Lemma 2.1.3 *The Gauss theorem – transformation of the volume integral into the surface integral*

$$\int_\Omega \nabla_i \sigma^{ij} d\Omega = \int_\Gamma n_i \sigma^{ij} d\Gamma \qquad (2.7)$$

where σ^{ij} and n_i represent the stress tensor and the surface normal components, respectively.

Using the Gauss theorem, the left hand side of equation (2.6) is transformed as follows

$$\begin{aligned}
\int_\Omega \nabla_i \sigma^{ij} \delta u_j d\Omega &= \int_\Omega \nabla_i (\sigma^{ij} \delta u_j) d\Omega - \int_\Omega \sigma^{ij} \nabla_i \delta u_j d\Omega \\
&= \int_\Gamma n_i (\sigma^{ij} \delta u_j) d\Gamma - \int_\Omega \sigma^{ij} \nabla_i \delta u_j d\Omega \\
&= \int_\Gamma \sigma^j_{(n)} \delta u_j d\Gamma - \int_\Omega \sigma^{ij} \nabla_i \delta u_j d\Omega,
\end{aligned} \qquad (2.8)$$

where the stress vector $\sigma^j_{(n)}$ results according to the Cauchy stress theorem in equation (2.4). Equation (2.8) can be further simplified by first considering that $\int_\Gamma \sigma^j_{(n)} \delta u_j d\Gamma$ is given only on a part of the whole surface Γ_σ:

$$\int_\Gamma \sigma^j_{(n)} \delta u_j d\Gamma = \int_{\Gamma_u} + \int_{\Gamma_\sigma} + \int_{\Gamma \backslash (\Gamma_u \cup \Gamma_\sigma)} = \int_{\Gamma_\sigma} \sigma^j_{(n)} \delta u_j d\Gamma,$$

because $\delta u^j = 0$ holds on Γ_u and $\sigma^j_{(n)} = 0$ holds on $\Gamma \backslash (\Gamma_u \cup \Gamma_\sigma)$.

The term $\int_\Omega \sigma^{ij} \nabla_i \delta u_j d\Omega$ can be transformed for small deformations using the Cauchy strain tensor and considering the symmetry property of σ^{ij} as

$$\int_\Omega \sigma^{ij} \nabla_i \delta u_j d\Omega = \int_\Omega \sigma^{ij} \delta \varepsilon_{ij} d\Omega. \qquad (2.9)$$

Thus, equation (2.6) finally is written as the following weak form

$$\underbrace{-\int_\Omega \sigma^{ij} \delta \epsilon_{ij} d\Omega}_{\delta W^i} + \underbrace{\int_{\Gamma_\sigma} \sigma^k_{(n)} \delta u_k d\Gamma}_{\delta W^{\Gamma_\sigma}} + \underbrace{\int_\Omega f^l \delta u_l d\sigma}_{\delta W^f} = 0 \qquad (2.10)$$

where δW^i represents the variation of the work of internal stresses, δW^{Γ_σ} represents the variation of the work of boundary forces $\sigma^k_{(n)}$ (the Neumann boundary conditions) and δW^f represents the variation of the work of volume forces f^l. During further solutions (analytical or numerical) special attention should be paid to Dirichlet boundary conditions.

General Formulation of a Contact Problem

Definition 2.1.4 *Equation (2.10) is well known in continuum mechanics as the variational equation for equilibrium, the weak form or the principle of virtual work.*

Remark 2.1.5 *One can see that, even though the problem has been formulated as the geometrically linear problem with linear elastic material, the consideration of contact leads to the nonlinearity, see Figure 2.1, because of the following:*

- *Contact conditions including both the non-penetration and stresses are not known.*
- *Contact conditions should be specified on the unknown contact area.*

The work of contact forces can be described by the term $\int_{\Gamma_\sigma} \sigma^k_{(n)} \delta u_k d\Gamma$, however, with the following unknowns: $\sigma^k_{(n)}$, δu_k and the surface Γ_σ.

2.2 Formulation of the Contact Part (Signorini's problem)

The first problem, historically very important for the development of computational contact mechanics, is the problem of contact with a rigid foundation, the so-called Signorini's problem (1933). Signorini's problem, shown schematically in Figure 2.2, studies the deformations of a body Ω, which potentially comes into contact with a rigid, fixed obstacle. Geometrically, contact conditions are formulated as follows: a point on the boundary is not allowed to penetrate the rigid obstacle. For the 2D case, discussed here for simplicity, this condition can be formulated using the position vectors r for the boundary of the deformable body ($\mathbf{r}_{ref}$ for the initial undeformed configuration) and the vector $\mathbf{r}_{rig}$ for the rigid obstacle, see Figure 2.2. In a 2D Cartesian coordinate system, these vectors are expressed as

$$\mathbf{r}_{rig} = \begin{Bmatrix} x \\ \Psi(x) \end{Bmatrix} \text{ and } \mathbf{r}_{ref} = \begin{Bmatrix} x \\ \Theta(x) \end{Bmatrix}. \quad (2.11)$$

For simplicity, we consider here surfaces for which parametrization in the form of equation (2.11) is possible. After deformation, the current position vector r is extended by the displacement vector as $\mathbf{u} = \begin{Bmatrix} u(x) \\ v(x) \end{Bmatrix}$

$$\mathbf{r} = \mathbf{r}_{ref} + \mathbf{u} = \begin{Bmatrix} x + u \\ \Theta + v \end{Bmatrix}. \quad (2.12)$$

The non-penetration contact condition (contact constraint) is formulated in the initial undeformed configuration as $\Theta(x)$ lays below $\Psi(x)$:

$$\Theta(x) \leq \Psi(x) \quad (2.13)$$

and is written for the current configuration as

$$\Theta(x) + v(x) \leq \Psi(x + u). \quad (2.14)$$

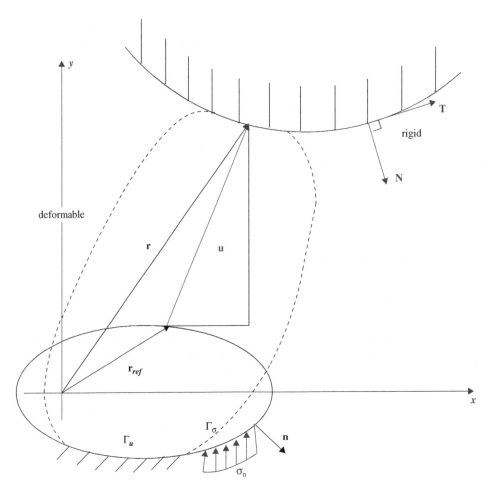

Figure 2.2 Contact with a rigid obstacle: Signorini's problem

After expanding $\Psi(x+u)$ into the Taylor series for only small deformations, equation (2.14) is written as the non-penetration condition for contacting bodies

$$\Theta(x) + v(x) \leq \Psi(x) + \Psi'(x)u(x)$$
$$\Theta - \Psi \leq -v + \Psi'u. \qquad (2.15)$$

This non-penetration condition can be expressed also using the unit normal vector n on the boundary of the obstacle. First, the tangent vector is calculated as the derivative of the vector $\mathbf{r}_{rig}$: $\mathrm{T} = \frac{d\mathbf{r}_{rig}}{dx} = \begin{pmatrix} 1 \\ \Psi'(x) \end{pmatrix}$. Then, the normal vector N is constructed to be orthogonal to the tangent vector T, thus, satisfying the orthogonality condition $\mathrm{T} \cdot \mathrm{N} = 0$. This is obtained in the form $\mathrm{N} = \begin{pmatrix} \Psi'(x) \\ -1 \end{pmatrix}$. Now, inequality (2.15) can be

General Formulation of a Contact Problem

written as the scalar product

$$\Theta - \Psi \leq (u\ v) \cdot \begin{pmatrix} \Psi' \\ -1 \end{pmatrix} = \mathbf{u} \cdot \mathbf{N}.$$

or with unit normal vector $\mathbf{n} = \frac{\mathbf{N}}{|\mathbf{N}|}$ as

$$\frac{\Theta - \Psi}{\sqrt{1 + (\Psi')^2}} \leq \mathbf{u} \cdot \mathbf{n}. \tag{2.16}$$

Geometrically, the scalar product $\mathbf{u} \cdot \mathbf{n}$ is describing the projection of a vector $\mathbf{u}$ onto the direction $\mathbf{n}$, or the normal component $u_{(n)}$ of the vector $\mathbf{u}$. One can see that the left part in equation (2.16) describes the distance along y-axis projected onto the normal direction $\mathbf{n}$ as well. This distance is called *the normalized initial gap*, however, with a minus sign, or the initial gap $-g$, or initial penetration in the normal direction. Using these notations, the inequality (2.16) is transformed as

$$-u_{(n)} - g \leq 0. \tag{2.17}$$

If the initial penetration is zero $g = 0$, that is bodies are initially in contact, then during the further deformation the following inequality should be satisfied

$$-u_{(n)} \leq 0. \tag{2.18}$$

Remark 2.2.1 *The geometrical interpretation of equation (2.18) is a normal component of the displacement vector $\mathbf{u}$ for the deformable body in the direction of the normal vector $\mathbf{n}$ of the rigid body should be more than zero during the deformation with contact. Now introducing a normal to the deformable body as $\mathbf{n}_{def}$ and taking into account that during the contact the normals coincide ($\mathbf{n} = -\mathbf{n}_{def}$), we can transform equation (2.18) into*

$$u_{(\mathbf{n}_{def})} \leq 0 \tag{2.19}$$

with more simple geometrical interpretation as a normal component of the displacement vector $\mathbf{u}$ to the contacting surface should be less than zero during the deformation with contact (penetration condition).

The condition in equation (2.19) is known as the non-penetration condition and appears to be the most important issue in computational contact mechanics.

This specific definition of the non-penetration condition correlates strongly with the stresses that are necessary to define in order to describe the contact conditions in the sense of the Karush–Kuhn–Tucker conditions. The normal force or stress vector $\sigma^{(n)}$ on the contact boundary can be split into the normal and tangential (shear) part:

$$\sigma^{(n)} = \sigma_n^{(n)} \mathbf{n} + \sigma_T^{(n)} \boldsymbol{\tau} \tag{2.20}$$

where $\boldsymbol{\tau}$ is the unit tangent vector, see Figure 2.3.

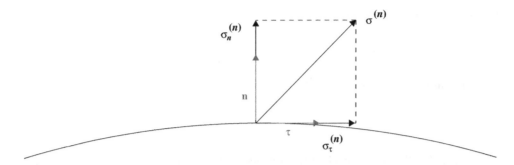

Figure 2.3 Split of the stress vector $\sigma^{(n)}$ into normal and tangential parts

Now considering only non-frictional problems defined as problems without tangential (shear) forces on the contact boundaries ($\sigma_{(n)}^\tau = 0$), the KKT conditions can be formulated for the deformed part ($\mathbf{n} \equiv \mathbf{n}_{def}$) as follows

$$\underbrace{\begin{array}{cc} \text{contact} & \text{no contact} \\ u_{(n)} - g = 0 \text{ and } \sigma_{(n)}^n < 0 & u_{(n)} - g < 0 \text{ and } \sigma_{(n)}^n = 0 \end{array}}$$

plus complimentary condition
$$(u_{(n)} - g)\sigma_{(n)}^n = 0.$$

In order to enforce the inequality conditions, the weak form in equation (2.10) is expanded by the virtual work carried out by the stress vector on the contacting surface. Thus, the Signorini problem is stated as

$$\delta W^i - \delta W^{\Gamma_\sigma} - \delta W^f + \delta W^c = 0$$

where the virtual contact work in the non-frictionless case is

$$\begin{aligned} \delta W^c &= \int_{\Gamma_c} \sigma_{(n)}^i \delta u_i d\Gamma \\ &= \int_{\Gamma_c} \boldsymbol{\sigma}^{(n)} \cdot \delta \mathbf{u} \, d\Gamma \\ &= \int_{\Gamma_c} \sigma_n^{(n)} (\mathbf{n} \cdot \delta \mathbf{u}) d\Gamma \\ &= \int_{\Gamma_c} \sigma_n^{(n)} \delta u_n d\Gamma \end{aligned} \qquad (2.21)$$

For more detail, see Kikuchi (1988).

Remark 2.2.2 *The most important achievement of Signorini's problem for further computational contact mechanics is that contact conditions such as non-penetration are formulated in the direction of the normal vector to the contacting body. Following further consideration, we will start to build computational algorithms assuming that this condition is already known.*

3

Differential Geometry

Since contact between bodies is naturally observed as geometrical interaction between surfaces and curves, the knowledge of principles of differential geometry is absolutely necessary for further developments of computational algorithms in contact mechanics. Differential geometry studies both geometrical properties of curves and surfaces and their differential properties. The differential geometry of curves (Sections 3.1 and 3.2) is necessary to build computational algorithms in the 2D space mostly considered in this book. The differential geometry of surfaces (Sections 3.3 and 3.4) is a general requirement to build computational algorithms in 3D space and can be skipped without losing the generality for most of the examples considered in Part II.

3.1 Curve and its Properties

A curve is a one-dimensional (1D) manifold in 3D space and can be defined in a vector form as

$$\boldsymbol{\rho} = \boldsymbol{\rho}(\xi),$$

where ξ is a parameter that can later be interpreted as a convective coordinate. We are concentrating on a curve in 2D, or just a plane curve. In order to study properties of the curve, an arc-length parameter s is introduced. This parameter describes the full length of the curve. Its differential is naturally defined via the scalar product

$$ds^2 = \left(\frac{\partial \boldsymbol{\rho}}{\partial \xi} d\xi \cdot \frac{\partial \boldsymbol{\rho}}{\partial \xi} d\xi\right) = \boldsymbol{\rho}_\xi \cdot \boldsymbol{\rho}_\xi (d\xi)^2 \qquad (3.1)$$

$$\text{or} \quad ds = \sqrt{\boldsymbol{\rho}_\xi \cdot \boldsymbol{\rho}_\xi}\, d\xi. \qquad (3.2)$$

A unit tangent vector $\boldsymbol{\tau}$ to the curve is defined as

$$\boldsymbol{\tau} = \frac{d\boldsymbol{\rho}}{ds} = \frac{\boldsymbol{\rho}_\xi}{\sqrt{\boldsymbol{\rho}_\xi \cdot \boldsymbol{\rho}_\xi}} \qquad (3.3)$$

Introduction to Computational Contact Mechanics: A Geometrical Approach, First Edition.
Alexander Konyukhov and Ridvan Izi.
© 2015 John Wiley & Sons, Ltd. Published 2015 by John Wiley & Sons, Ltd.
Companion Website: www.wiley.com/go/Konyukhov

In order to study properties of the curve the Frenet formulas are used. These formulas are defined as a set of derivatives from the unit curve vectors with respect to the arc-length parameter s. First, we define a derivative of τ for which we need the following lemma.

Lemma 3.1.1 *The derivative of a unit vector is orthogonal to this vector.*

Proof. For any unit vector τ possessing the length $\tau \cdot \tau = 1$, the derivative with respect to a variable s is written as

$$\frac{d}{ds}(\tau \cdot \tau) = \frac{d\tau}{ds} \cdot \tau + \tau \cdot \frac{d\tau}{ds} = 0$$

$$\Rightarrow \frac{d\tau}{ds} \cdot \tau = 0.$$

The last scalar product of course confirms the orthogonality of vectors τ and $\frac{d\tau}{ds}$.

Considering the derivation then geometrically as

$$\frac{d\tau}{ds} = \lim_{\Delta s \to 0} \frac{\tau(s + \Delta s) - \tau(s)}{\Delta s}, \tag{3.4}$$

see Figure 3.1, one can clearly observe that a vector $\Delta \tau$ (and later $\frac{d\tau}{ds}$) is pointing into the convex part of the curve. Namely, in this direction we define the unit normal vector, ν. Two vectors τ and ν define a coordinate system called a Frenet frame. Due

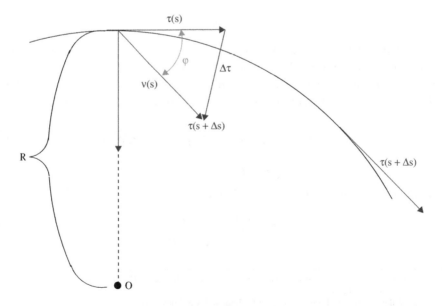

Figure 3.1 Definition of the Frenet frame and the center of curvature

to Lemma 3.1.1, the derivative $\frac{d\tau}{ds}$ is orthogonal to τ and, therefore, is proportional to the normal vector ν as

$$\frac{d\tau}{ds} = k\nu \tag{3.5}$$

Equation (3.5) is also known as the first Frenet formula.

Thus, the normal vector ν is defined as a normed derivative of the unit tangent

$$\nu = \frac{\frac{d\tau}{ds}}{\left|\frac{d\tau}{ds}\right|} \tag{3.6}$$

Coefficient k is called the curvature of a curve. We will study its property by testing it on the simple curve – a circle, see Figure 3.2. The curvature k is defined via a scalar product of equation (3.5) with a vector ν

$$k = \frac{d\tau}{ds} \cdot \nu = \left|\frac{d\tau}{ds}\right| \tag{3.7}$$

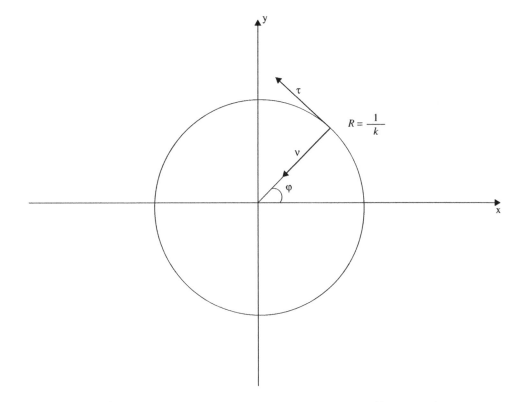

Figure 3.2 A circle as a curve with constant curvature and its Frenet frame

which is always positive due to definition of ν. In order to construct a formula convenient for computation we exploit equation (3.2) and definition of τ in equation (3.3)

$$\begin{aligned}
k &= \frac{d\tau}{ds} \cdot \nu \\
&= \frac{d}{d\xi}\left(\frac{d\rho}{ds}\right)\frac{d\xi}{ds} \cdot \nu \\
&= \frac{d}{d\xi}\left(\frac{d\rho}{d\xi}\frac{d\xi}{ds}\right)\frac{d\xi}{ds} \cdot \nu \\
&= \frac{d^2\rho}{(d\xi)^2}\left(\frac{d\xi}{ds}\right)^2 \cdot \nu + \underbrace{\frac{d\rho}{d\xi}\frac{d^2\xi}{d\xi ds}\frac{d\xi}{ds} \cdot \nu}_{=\,0,\ \text{because}\ \tau \cdot \nu = 0} \\
&= \frac{\frac{d^2\rho}{(d\xi)^2} \cdot \nu}{\left(\frac{ds}{d\xi}\right)^2} = \frac{\rho_{\xi\xi} \cdot \nu}{\rho_\xi \cdot \rho_\xi}.
\end{aligned}$$

During the final step, equation (3.2) is again taken into account. Finally the equation for curvature is

$$k = \frac{\rho_{\xi\xi} \cdot \nu}{\rho_\xi \cdot \rho_\xi}. \qquad (3.8)$$

Without knowing the normal vector ν, the curvature can be computed as the length of the vector $\frac{d\tau}{ds}$ from equation (3.5), therefore,

$$k = \left|\frac{d\tau}{ds}\right|. \qquad (3.9)$$

The last equation (3.9) allows the simple geometrical interpretation of the curvature – this is the derivative of the angle φ during motion of the unit tangent τ, see Figure 3.1, with respect to the arc-length s:

$$k = \pm\frac{d\varphi}{ds}, \qquad (3.10)$$

where the "plus" or "minus" sign depends on the convention of positive rotation.

3.1.1 Example: Circle and its Properties

The simplest curve geometry – a circle – is taken to study the geometrical meaning of the curvature. Parametrization of a circle on the plane in the form $\rho = \rho(\varphi)$ is written as

$$\begin{cases} x = R\cos\varphi \\ y = R\sin\varphi \end{cases}$$

Differential Geometry

Derivatives with respect to parameter φ (as a convective coordinate ξ)

$$\frac{d\rho}{d\varphi} : \begin{cases} \dfrac{dx}{d\varphi} = -R\sin\varphi \\ \dfrac{dy}{d\varphi} = R\cos\varphi \end{cases}$$

define a tangent vector. An arc-length differential in equation (3.2) is calculated as

$$ds = \sqrt{\rho_\varphi \cdot \rho_\varphi}\, d\varphi = \sqrt{R^2\sin^2\varphi + R^2\cos^2\varphi}\, d\varphi = R\, d\varphi, \qquad (3.11)$$

then the Jacobian for transformation of variables is

$$\frac{d\varphi}{ds} = \frac{1}{R}.$$

A unit tangent vector follows equation (3.3):

$$\tau : \begin{cases} \tau_x = \dfrac{dx}{ds} = \dfrac{dx}{d\varphi}\dfrac{d\varphi}{ds} = -\sin\varphi \\ \tau_y = \dfrac{dy}{ds} = \dfrac{dy}{d\varphi}\dfrac{d\varphi}{ds} = \cos\varphi \end{cases}$$

Its derivative with respect to s:

$$\frac{d\tau}{ds} : \begin{cases} \dfrac{d\tau_x}{ds} = \dfrac{d\tau_x}{d\varphi}\dfrac{d\varphi}{ds} = -\dfrac{\cos\varphi}{R} \\ \dfrac{d\tau_y}{ds} = \dfrac{d\tau_y}{d\varphi}\dfrac{d\varphi}{ds} = -\dfrac{\sin\varphi}{R} \end{cases}$$

The absolute value gives us the curvature in equation (3.9)

$$\left|\frac{d\tau}{ds}\right| = \sqrt{\frac{\cos^2\varphi + \sin^2\varphi}{R^2}} = \frac{1}{R}. \qquad (3.12)$$

The normal vector ν is defined as a normed derivative in equation (3.6)

$$\nu = \frac{\frac{d\tau}{ds}}{\left|\frac{d\tau}{ds}\right|} = \begin{Bmatrix} -\cos\varphi \\ -\sin\varphi \end{Bmatrix}$$

and the curvature in equation (3.8) is again recovered

$$k = \frac{d\tau}{ds} \cdot \nu = \frac{1}{R}.$$

Now the geometrical meaning of the curvature is clear – this is the inverse value of the radius.

Definition 3.1.2 *The inverse value of the curvature is called the radius of curvature.*

$$R = \frac{1}{k}$$

For any arbitrary curve, a point O in the direction of the normal ν at the distance R from the curve can be defined, see Figure 3.1. This point is called the *center of curvature*.

3.2 Frenet Formulas in 2D

Frenet formulas are defining derivatives of the unit vectors τ and ν with respect to the arc-length s. Together with the first Frenet formula in equation (3.5) we define the second formula determining the derivative of ν. Since ν is a unit vector (using Lemma 3.1.1) we again recover the orthogonality of the derivative of unit vector to this unit vector, and therefore, proportionality to the vector τ:

$$\frac{d\nu}{ds} = c\tau \tag{3.13}$$

Before the next step, we need to obtain an auxiliary equation by taking the derivative of two orthogonal vectors τ and ν

$$\frac{d}{ds}(\tau \cdot \nu) = \frac{d\tau}{ds} \cdot \nu + \tau \cdot \frac{d\nu}{ds} = 0 \tag{3.14}$$

Now we can determine c. Multiplying equation (3.13) with the vector τ as a scalar product and taking into account the relationship given in equation (3.14), the following result is obtained

$$c(\tau \cdot \tau) = \frac{d\nu}{ds} \cdot \tau = -\nu \cdot \frac{d\tau}{ds} \tag{3.15}$$

For the last step to determine a coefficient c, the first Frenet formula equation (3.5) is taken into account.

$$c = -\nu \cdot k\nu = -k \tag{3.16}$$

Thus to summarize, we obtain the full set of Frenet formulas as

$$\begin{cases} \dfrac{d\tau}{ds} = k\nu \\ \dfrac{d\nu}{ds} = -k\tau \end{cases} \tag{3.17}$$

The Frenet formulas equation (3.17) is used to describe any kinematic properties within a non-inertial coordinate system defined by the basis vectors τ and ν (Frenet frame).

3.3 Description of Surfaces by Gauss Coordinates

A surface is a two-dimensional (2D) manifold in 3D Cartesian space and naturally understood to be a boundary of a 3D body. The surface can be defined as a specification of three functions for Cartesian coordinates x,y,z depending on two parameters ξ^1, ξ^2:

$$\left\{\begin{array}{l} x = x(\xi^1, \xi^2) \\ y = y(\xi^1, \xi^2) \\ z = z(\xi^1, \xi^2) \end{array}\right\} \tag{3.18}$$

where ξ^1, ξ^2 are local convective coordinates defining the surface coordinate lines, see Figure 3.3. In a vector form, it is written in short as

$$\boldsymbol{\rho} = \boldsymbol{\rho}(\xi^1, \xi^2). \tag{3.19}$$

Coordinates ξ^1, ξ^2 are called *contravariant coordinates* or *Gaussian coordinates*.

3.3.1 Tangent and Normal Vectors: Surface Coordinate System

In order to introduce a coordinate system on a surface, two tangent vectors are computed as derivatives with respect to contravariant variables

$$\boldsymbol{\rho}_1 = \frac{\partial \boldsymbol{\rho}}{\partial \xi^1}, \quad \boldsymbol{\rho}_2 = \frac{\partial \boldsymbol{\rho}}{\partial \xi^2}. \tag{3.20}$$

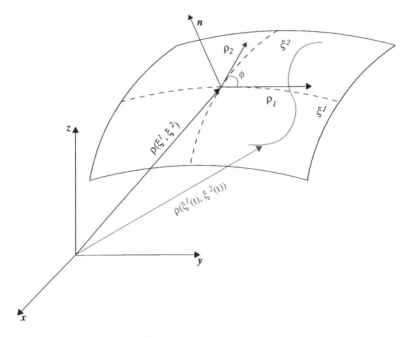

Figure 3.3 Surface in space: definition of a surface coordinate system $(\boldsymbol{\rho}_1, \boldsymbol{\rho}_2, \mathbf{n})$. A curve on the surface is defined as $\boldsymbol{\rho}(\xi^1(t), \xi^2(t))$

They are called covariant tangent vectors of the surface. The normal vector $\mathbf{N}$ to the surface is defined via the cross product of tangent vectors

$$\mathbf{N} = \boldsymbol{\rho}_1 \times \boldsymbol{\rho}_2. \tag{3.21}$$

Cartesian coordinates of the vector $\mathbf{N}$ are defined in the form of determinant with Cartesian unit vectors $\mathbf{i}, \mathbf{j}, \mathbf{k}$

$$\mathbf{N} = \begin{bmatrix} \mathbf{i} & \mathbf{j} & \mathbf{k} \\ \frac{\partial x}{\partial \xi^1} & \frac{\partial y}{\partial \xi^1} & \frac{\partial z}{\partial \xi^1} \\ \frac{\partial x}{\partial \xi^2} & \frac{\partial y}{\partial \xi^2} & \frac{\partial z}{\partial \xi^2} \end{bmatrix} = \mathbf{i} \begin{bmatrix} \frac{\partial y}{\partial \xi^1} & \frac{\partial z}{\partial \xi^1} \\ \frac{\partial y}{\partial \xi^2} & \frac{\partial z}{\partial \xi^2} \end{bmatrix} - \mathbf{j} \begin{bmatrix} \frac{\partial x}{\partial \xi^1} & \frac{\partial z}{\partial \xi^1} \\ \frac{\partial x}{\partial \xi^2} & \frac{\partial z}{\partial \xi^2} \end{bmatrix} + \mathbf{k} \begin{bmatrix} \frac{\partial x}{\partial \xi^1} & \frac{\partial y}{\partial \xi^1} \\ \frac{\partial x}{\partial \xi^2} & \frac{\partial y}{\partial \xi^2} \end{bmatrix}.$$

$$\tag{3.22}$$

For further analysis a unit normal $\mathbf{n}$ is employed, therefore, $\mathbf{N}$ is normalized as:

$$\mathbf{n} = \frac{\boldsymbol{\rho}_1 \times \boldsymbol{\rho}_2}{|\boldsymbol{\rho}_1 \times \boldsymbol{\rho}_2|}$$

Three basis vectors, $\boldsymbol{\rho}_1, \boldsymbol{\rho}_2, \mathbf{n}$, form a surface coordinate system.

3.3.2 Basis Vectors: Metric Tensor and its Applications

It is important to distinguish between contravariant components with upper indexes (as introduced for coordinates in equation (3.18) and covariant ones with lower indexes (as it is written in equation (3.20). Duality of covariant and contravariant components is disclosed in the calculation of a length differential on the surface

$$ds^2 = d\boldsymbol{\rho} \cdot d\boldsymbol{\rho} = \left(\frac{\partial \boldsymbol{\rho}}{\partial \xi^1} d\xi^1 + \frac{\partial \boldsymbol{\rho}}{\partial \xi^2} d\xi^2 \right) \cdot \left(\frac{\partial \boldsymbol{\rho}}{\partial \xi^1} d\xi^1 + \frac{\partial \boldsymbol{\rho}}{\partial \xi^2} d\xi^2 \right) =$$

$$= \frac{\partial \boldsymbol{\rho}}{\partial \xi^i} d\xi^i \cdot \frac{\partial \boldsymbol{\rho}}{\partial \xi^j} d\xi^j = \left(\frac{\partial \boldsymbol{\rho}}{\partial \xi^i} \cdot \frac{\partial \boldsymbol{\rho}}{\partial \xi^j} \right) d\xi^i d\xi^j = (\boldsymbol{\rho}_i \cdot \boldsymbol{\rho}_j) d\xi^i d\xi^j = a_{ij} d\xi^i d\xi^j \tag{3.23}$$

Here, the metric tensor is introduced with components $a_{ij} = \boldsymbol{\rho}_i \cdot \boldsymbol{\rho}_j$.

The form $a_{ij} d\xi^i d\xi^j$ is called the *first fundamental form of a surface*. Its components are written in a covariant form (lower indexes). Thus, the metric tensor is given by the matrix

$$[a_{ij}] = \begin{bmatrix} a_{11} & a_{12} \\ a_{21} & a_{22} \end{bmatrix}. \tag{3.24}$$

Equation (3.23) discloses the duality: the sum over lower and upper indexes i, j leads to an invariant product (vector, scalar) – in the current case to a scalar value – the length.

3.3.2.1 Metric Properties of the Surface

Components of the metric tensor a_{ij} are used to define the metric properties of a surface, namely to measure

- the length of a curve laying on a surface;
- the area of a surface;
- the angle between two tangent vectors.

3.3.2.2 The Length of a Curve Laying on a Surface

Consider a curve which belongs to a surface with metric components a_{ij}, written in the form

$$\rho = \rho(\xi^1(t), \xi^2(t)) \tag{3.25}$$

Since the curve belongs to a surface, its equation is written as a part of the surface equation, see Figure 3.3. The inner equations defining a curve on the surface are written via convective coordinates as

$$\xi^1 = \xi^1(t); \quad \xi^2 = \xi^2(t), \tag{3.26}$$

while equation (3.25) describes the surface with variables ξ^1, ξ^2, as in equation (3.19). The length L, when the parameter t changes from t_A to t_B, is computed as

$$L = \int_{t_A}^{t_B} \sqrt{\frac{d\rho}{dt} \cdot \frac{d\rho}{dt}} \, dt = \int_{t_A}^{t_B} \sqrt{\frac{\partial \rho}{\partial \xi^i} \frac{d\xi^i}{dt} \cdot \frac{\partial \rho}{\partial \xi^j} \frac{d\xi^j}{dt}} \, dt =$$

$$= \int_{t_A}^{t_B} \sqrt{(\rho_i \cdot \rho_j) \dot{\xi}^i \dot{\xi}^j} \, dt = \int_{t_A}^{t_B} \sqrt{a_{ij} \dot{\xi}^i \dot{\xi}^j} \, dt. \tag{3.27}$$

3.3.2.3 Angle between Two Curves

Consider two coordinate lines as an example of two curves belonging to the surface. The angle between them is defined as an angle between tangent vectors to those curves at the point of intersection, see Figure 3.3. The cosine is computed as

$$\cos \phi = \frac{(\rho_1 \cdot \rho_2)}{\sqrt{\rho_1 \cdot \rho_1} \sqrt{\rho_2 \cdot \rho_2}} = \frac{a_{12}}{\sqrt{a_{11}} \sqrt{a_{22}}}. \tag{3.28}$$

The geometrical interpretation of covariant components of the metric tensor is as follows:

1. diagonal components a_{11}, a_{22} define the length of the corresponding basis vector ρ_1, ρ_2

$$a_{ii} = (\rho_i \cdot \rho_i) = |\rho_i|^2.$$

2. non-diagonal component a_{12} is used for the definition of the cosine between ρ_1 and ρ_2, see equation (3.28).

3.3.2.4 Area of Surface

From analytical geometry it is known that the cross product between ρ_1 and ρ_2 is a vector, **N**, orthogonal to them with the length proportional to the area spanned by those vectors. We transform the square of the length as follows

$$|\mathbf{N}|^2 = |\rho_1 \times \rho_2|^2 = |\rho_1|^2|\rho_2|^2 \sin^2\phi =$$
$$= |\rho_1|^2|\rho_2|^2(1 - \cos^2\phi) = |\rho_1|^2|\rho_2|^2 - |\rho_1|^2|\rho_2|^2\cos^2\phi =$$
$$= a_{11}a_{22} - a_{12}^2 = det(a_{ij}). \tag{3.29}$$

During the last transformation, the geometrical properties of the scalar (cosine) and the cross (sine) product have been employed. For an arbitrary area, see Figure 3.4, an integral over the surface must be computed

$$S = \int_S |\rho_1 \times \rho_2| \, d\xi^1 d\xi^2 = \int_S \sqrt{a_{11}a_{22} - a_{12}^2} \, d\xi^1 d\xi^2. \tag{3.30}$$

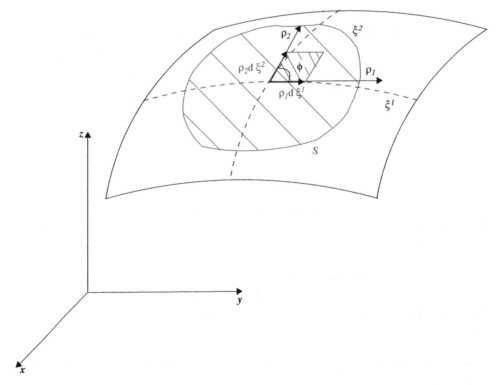

Figure 3.4 Calculation of a surface area S. Infinitesimal area $dS = |\mathbf{N}|d\xi^1 d\xi^2 = |\rho_1 \times \rho_2|d\xi^1 d\xi^2$

3.3.3 Relationships between Co- and Contravariant Basis Vectors

The covariant components of the metric tensor are given by the matrix in equation (3.24). The inverse matrix gives us the contravariant components a^{ij}

$$[a_{ij}]^{-1} = a^{ij} = \begin{pmatrix} a^{11} & a^{12} \\ a^{21} & a^{22} \end{pmatrix} \tag{3.31}$$

The relation between co- and contravariant components is disclosed by the Kronecker delta, which defines the mixed components of the metric tensor a_i^k.

$$a_{ij}a^{jk} = a_i^k = \delta_i^k \rightarrow \begin{bmatrix} 1 & 0 \\ 0 & 1 \end{bmatrix}. \tag{3.32}$$

Mixed components a_i^j are used to define the contravariant basis vectors ρ^j via the scalar product as

$$(\rho_i \cdot \rho^j) = \delta_i^j = a_i^j \tag{3.33}$$

Thus, the contravariant basis can be formed as

$$\rho^i = a^{ij}\rho_j = a^{i1}\rho_1 + a^{i2}\rho_2, \tag{3.34}$$

satisfying equation (3.33).

The geometric interpretation of the contravariant basis is given on the tangent plane, see Figure 3.5

1. ρ_1 is orthogonal to ρ^2 (1⇌2);
2. the angle between ρ_1 and ρ^1 (1⇌2) is less than 90° (acute angle) due to the cosine property in equation (3.28).

$$\rho_1 \cdot \rho^1 = 1 \rightarrow \cos(\widehat{\rho_1\rho^1}) > 0 \rightarrow -90° < \widehat{\rho_1\rho^1} < 90°. \tag{3.35}$$

Contravariant vectors ρ^i belong to the surface, therefore, they are represented via covariant ones as

$$\rho^i = a^{ij}\rho_j \tag{3.36}$$

Scalar product equation (3.36), with ρ_k taking into account equation (3.33), recovers that

$$\rho^i \cdot \rho_k = a^{ij}\rho_j \cdot \rho_k \Rightarrow \delta_k^i = a^{ij}a_{jk} \tag{3.37}$$

which proves that a^{ij} are contravariant components as defined in equation (3.31). However, they can be computed as a scalar product of contravariant basis vectors after taking the scalar product equation (3.36) with ρ^k as

$$a^{ij} = \rho^i \cdot \rho^j \tag{3.38}$$

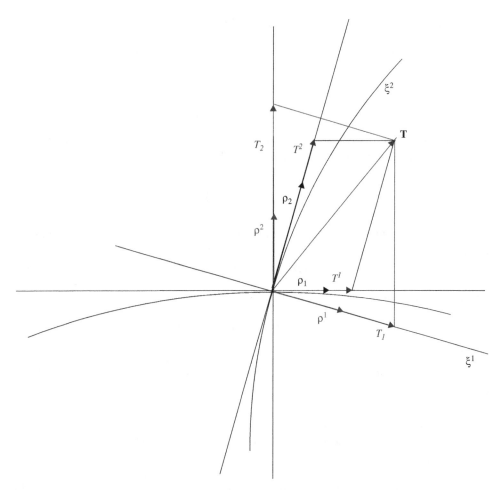

Figure 3.5 Covariant ρ_1, ρ_2 and contravariant ρ^1, ρ^2 coordinate vectors. Representation of a vector via co- and contravariant components

Finally, in the same manner, we obtain the representation of covariant vectors via contravariant ones

$$\rho_i = a_{ij}\rho^j. \tag{3.39}$$

Definition 3.3.1 *Formally, equation (3.36) makes it look like the sum of the covariant (lower j index object) vector with contravariant metric components (double upper j, i index object) leads to the contravariant vector (upper i index object). Therefore, the rule in equation (3.36) is called the* rising index rule *and the rule in equation (3.39), the lowering index rule.*

3.3.4 Co- and Contravariant Representation of a Vector on a Surface

A vector $\mathbf{T}$ belongs to a surface if $\mathbf{T}$ belongs to a tangent plane of the surface. The vector can be represented via covariant vectors ρ_i with contravariant components T^i

as a contravariant vector

$$\mathbf{T} = T^1 \boldsymbol{\rho}_1 + T^2 \boldsymbol{\rho}_2 = T^i \boldsymbol{\rho}_i \qquad (3.40)$$

as well as via contravariant vectors $\boldsymbol{\rho}^i$ with covariant components T_i as a covariant vector

$$\mathbf{T} = T_1 \boldsymbol{\rho}^1 + T_2 \boldsymbol{\rho}^2 = T_i \boldsymbol{\rho}^i. \qquad (3.41)$$

The *absolute value* of the vector, or its length, is computed via contravariant components as

$$|\mathbf{T}|^2 = \mathbf{T} \cdot \mathbf{T} = T^i \boldsymbol{\rho}_i \cdot T^j \boldsymbol{\rho}_j = T^i T^j (\boldsymbol{\rho}_i \cdot \boldsymbol{\rho}_j) = \qquad (3.42)$$

$$= T^i T^j a_{ij} = T^1 T^1 a_{11} + 2T^1 T^2 a_{12} + T^2 T^2 a_{22} \qquad (3.43)$$

or via covariant components as

$$|\mathbf{T}|^2 = T_i T_j a^{ij}. \qquad (3.44)$$

Relationships between co- and contravariant components are subjected to the index rising, index lowing rules discussed in Definition 3.3.1, such as index raising $T_i a^{ij} = T^j$ and index lowering $T^j a_{ij} = T_i$.

3.3.5 Curvature Tensor and Structure of the Surface

Besides the metric properties (angles, length) given by the metric tensor, there are other essential properties for the surface, such as:

1. curvatures defining a surface as a curved structure in 3D Cartesian space,
2. curvatures defining inner properties of a surface that are invariants to the deformation of the surface. Thus, for example a rolled (or developing surface) sheet can represent a cylinder.

Computing second derivatives of the surface vector

$$\boldsymbol{\rho}_{ij} = \frac{\partial \boldsymbol{\rho}}{\partial \xi^i \partial \xi^j} \qquad (3.45)$$

covariant components of a curvature tensor $\mathbf{H}$ are introduced as

$$h_{ij} = (\mathbf{n} \cdot \boldsymbol{\rho}_{ij}). \qquad (3.46)$$

These components are forming the symmetric matrix for any two times differentiable point of the surface. The tensor $\mathbf{H}$ is formulated in the surface basis in various forms as:

$$\mathbf{H} = h_{ij} \boldsymbol{\rho}^i \otimes \boldsymbol{\rho}^j = h_{i \cdot}^{\cdot j} \boldsymbol{\rho}^i \otimes \boldsymbol{\rho}_j = h_{ij} \boldsymbol{\rho}^i \otimes \boldsymbol{\rho}^j. \qquad (3.47)$$

3.3.5.1 Analysis of the Curvature Tensor: Generalized Eigenvalue Problem

Definition 3.3.2 *The generalized eigenvalue problem for the surface tensor* $\mathbf{H}$ *is defined as: "Find an eigenvector* $\mathbf{e}$ *which belongs to the surface and an eigenvalue* k *such that":*

$$\mathbf{H}\mathbf{e} + k\mathbf{e} = 0. \tag{3.48}$$

Expressing equation (3.48) in contravariant basis vectors

$$h_{ij}\rho^i \otimes \rho^j \cdot e_k \rho^k + k e_k \rho^k = 0, \tag{3.49}$$

performing a scalar product in the first term and changing index k into i in the second term we obtain

$$h_{ij}\rho^i e_k a^{jk} + k e_i \rho^i = 0. \tag{3.50}$$

Thus follows

$$(h_{ij} e_k a^{jk} + k e_i)\rho^i = 0 \tag{3.51}$$

and the homogenous system of equations with unknowns e_i:

$$h_{ij} e_k a^{jk} + k e_i = 0 \tag{3.52}$$

or again changing k into n and using mixed components of the metric tensor as Kronecker-delta

$$h_i^n e_n + k a_i^n e_n = 0. \tag{3.53}$$

The homogenous system (3.53) is resolved via components e_n only if its determinant is zero:

$$det(h_i^n + k a_i^n) = 0. \tag{3.54}$$

If the system (3.54) is written via contravariant components e^n then we obtain

$$det(h_{ij} + k a_{ij}) = 0. \tag{3.55}$$

The solution of this square equation gives us two roots k_1, k_2 called the principal curvatures of the surface.

The following invariants can be constructed for the curvature tensor:

1. Gaussian curvature

$$K = k_1 \cdot k_2 = \frac{det(h_{ij})}{det(a_{ij})}.$$

2. The mean curvature

$$H = \frac{k_1 + k_2}{2} = \frac{1}{2}\frac{a_{22}h_{11} - 2a_{12}h_{12} + a_{11}h_{22}}{a_{11}a_{22} - a_{12}^2}.$$

Differential Geometry

The values of principal curvatures k_1, k_2 and the Gaussian curvature K allow us to study the local structure of the surface surrounding a point. The local structure of the surface is summarized in the table here. The local structure is named after the names of simple geometrical surfaces, namely, the surface surrounding the point looks locally like those surfaces, except the flat point for which further advanced analysis is necessary.

Gaussian curvature	Local structure
$K = k_1 k_2 > 0$	elliptic point
$K = k_1 k_2 < 0$	hyperbolic point
$K = 0, k_1 = 0, k_2 \neq 0 \quad 1 \rightleftarrows 2$	cylindrical point
$K = 0, k_1 = 0, k_2 = 0$ at point	flat point
$k_1 = 0, k_2 = 0$ at any point	plane

3.4 Differential Properties of Surfaces

First, it is necessary to calculate derivatives of surface vectors ρ_i and normal $\mathbf{n}$, in order to calculate derivatives of some objects (vector, tensor) given in the surface coordinate system. A set of formulas, called the Weingarten and the Gauss–Codazzi formulas, is used for these operations.

3.4.1 The Weingarten Formula

The Weingarten formula defines derivatives of the normal vector $\mathbf{n}$ with respect to convective (Gaussian) coordinates ξ^j. The derivative $\frac{\partial \mathbf{n}}{\partial \xi^i}$ is orthogonal to $\mathbf{n}$, because the normal vector $\mathbf{n}$ is a unit vector (see Lemma 3.1.1) and, therefore, is expressed by ρ_j

$$\frac{\partial \mathbf{n}}{\partial \xi^i} = \mathbf{n}_i = c_i^k \rho_k. \tag{3.56}$$

First, we consider the orthogonality of ρ_i and $\mathbf{n}$

$$\rho_i \cdot \mathbf{n} = 0, \tag{3.57}$$

and, taking then derivatives with respect to ξ^j, we obtain

$$(\rho_{ij} \cdot \mathbf{n}) + (\rho_i \cdot \mathbf{n}_j) = 0$$

The first term defines components of the curvature tensor h_{ij}

$$h_{ij} + \rho_i \cdot \mathbf{n}_j = 0. \tag{3.58}$$

Now we take the scalar product of equation (3.56) with ρ_j

$$c_i^k (\rho_k \cdot \rho_j) = (\mathbf{n}_i \cdot \rho_j)$$

and, taking into account equation (3.58), we obtain

$$c_i^k a_{kj} = (\mathbf{n}_i \cdot \boldsymbol{\rho}_j) = -h_{ij}.$$

Thus, the components c_i^k are defined as

$$c_i^k = -h_{ij}a^{jk} = -h_i^k \qquad (3.59)$$

Finally, equation (3.56) has the following form

$$\mathbf{n}_i = -h_i^k \boldsymbol{\rho}_k = -h_{ij}a^{jk}\boldsymbol{\rho}_k, \qquad (3.60)$$

which is called the Weingarten formula.

3.4.2 The Gauss–Codazzi Formula

The Gauss–Codazzi formulas define a set of derivatives of surface vectors $\boldsymbol{\rho}_i$ with respect to convective (Gaussian) coordinates ξ^1, ξ^2. Derivatives of coordinate vectors $\boldsymbol{\rho}_i$ are not obviously laying in the tangent plane and, therefore, they are expressed via both the surface $\boldsymbol{\rho}_k$ and the normal $\mathbf{n}$ vectors

$$\frac{\partial \boldsymbol{\rho}_i}{\partial \xi^j} = \boldsymbol{\rho}_{ij} = \Gamma_{ij}^k \boldsymbol{\rho}_k + h_{ij}\mathbf{n}. \qquad (3.61)$$

Γ_{ij}^k symbols are called the *Christoffel symbols for the surface*. In order to compute them, first, the scalar product of equation (3.61) with contravariant vector $\boldsymbol{\rho}^n$ is taken:

$$(\boldsymbol{\rho}_{ij} \cdot \boldsymbol{\rho}^n) = \Gamma_{ij}^k (\boldsymbol{\rho}_k \cdot \boldsymbol{\rho}^n) + h_{ij}(\mathbf{n} \cdot \boldsymbol{\rho}^n) \qquad (3.62)$$

Since $\boldsymbol{\rho}^i$ as well as $\boldsymbol{\rho}_i$ are always laying in the tangent plane we have $\mathbf{n} \cdot \boldsymbol{\rho}^n = 0$, and the Christoffel symbols are computed as

$$\Gamma_{ij}^n = (\boldsymbol{\rho}^n \cdot \boldsymbol{\rho}_{ij}) = a^{nk}(\boldsymbol{\rho}_k \cdot \boldsymbol{\rho}_{ij}). \qquad (3.63)$$

The scalar product of equation (3.61) with the normal vector $\mathbf{n}$ recovers the curvature tensor components h_{ij}:

$$(\boldsymbol{\rho}_{ij} \cdot \mathbf{n}) = \Gamma_{ij}^k (\boldsymbol{\rho}_k \cdot \mathbf{n}) + h_{ij}(\mathbf{n} \cdot \mathbf{n}) = h_{ij} \qquad (3.64)$$

The formula in equation (3.61) is called the Gauss–Codazzi formula.

3.4.3 Covariant Derivatives on the Surface

In order to take derivatives of an object given a curvilinear coordinate system one must take into account the derivatives of coordinate vectors as well. Consider a time-dependent vector $\mathbf{T}$ belonging to a surface $\boldsymbol{\rho}(\xi^1, \xi^2)$.

$$\mathbf{T}(\xi^1, \xi^2, t) = T^i(\xi^1, \xi^2, t)\boldsymbol{\rho}_i(\xi^1, \xi^2) \qquad (3.65)$$

Differential Geometry

We exclude the motion of the surface and focus only on changing **u** on the surface. The assumption of $\frac{\partial \rho_i}{\partial t} = 0$ excludes of the surface motion. A full time derivative gives us

$$\frac{d\mathbf{T}}{dt} = \frac{d(T^i)\rho_i}{dt} + T^i \frac{\partial \rho_i}{\partial t} \tag{3.66}$$

$$= \left(\frac{\partial T^i}{\partial t} + \frac{\partial T^i}{\partial \xi^j} \frac{d\xi^j}{dt} \right) \rho_i + T^i \frac{\partial \rho_i}{\partial \xi^j} \frac{d\xi^j}{dt}$$

$$= \frac{\partial T^i}{\partial t} \rho_i + \frac{\partial T^i}{\partial \xi^j} \dot\xi^j \rho_i + (\Gamma^k_{ij}\rho_k + h_{ij}\mathbf{n}) T^i \dot\xi^j$$

$$= \left[\frac{\partial T^i}{\partial t} + \left(\frac{\partial T^i}{\partial \xi^j} + \Gamma^i_{kj} u^k \right) \dot\xi^j \right] \rho_i + h_{ij} T^i \dot\xi^j \mathbf{n},$$

where the Gauss–Codazzi formula in equation (3.61) has been used.
Denote the following components as

$$\nabla_j T^i = \frac{\partial T^i}{\partial \xi^j} + \Gamma^i_{kj} T^k. \tag{3.67}$$

They are called covariant derivatives of contravariant components T^i. Projection of the vector $\frac{d\mathbf{u}}{dt}$ on the surface gives

$$\frac{dT^i}{dt} = \frac{\partial T^i}{\partial t} + \nabla_j T^i \dot\xi^j = \frac{\partial T^i}{\partial t} + \left(\frac{\partial T^i}{\partial \xi^j} + \Gamma^i_{kj} T^k \right) \dot\xi^j. \tag{3.68}$$

Equation (3.68) defines *the full time derivative of the contravariant vector in a covariant form*. If the vector is defined via covariant components as $\mathbf{T} = T_i \rho^i$, then its derivative $\frac{d\mathbf{T}}{dt}$ in the covariant form gives the components

$$\frac{dT_i}{dt} = \frac{\partial T_i}{\partial t} + \left(\frac{\partial T_i}{\partial \xi^j} - \Gamma^k_{ij} T_k \right) \dot\xi^j \tag{3.69}$$

which are called *covariant derivatives of covariant components*.

Definition 3.4.1 *Here, the Nabla operator ∇_i in the arbitrary curvilinear coordinate system in the covariant form is used, see comparison with the Definition 2.1.1.*

3.4.4 Example: Geometrical Analysis of a Cylindrical Surface

A cylindrical surface is obtained as a surface in the cylindrical coordinate system (r, φ, z) by setting the radial coordinate $r = R$ as the radius of the cylinder, see Figure 3.6:

$$\begin{cases} x = R\cos\varphi = R\cos\xi^1 \\ y = R\sin\varphi = R\sin\xi^1, \quad \varphi \in [0, 2\pi], \quad z \in (-\infty, +\infty) \\ z = \xi^2 \end{cases} \tag{3.70}$$

with local coordinates $\xi^1 \equiv \varphi$, $\xi^2 \equiv z$

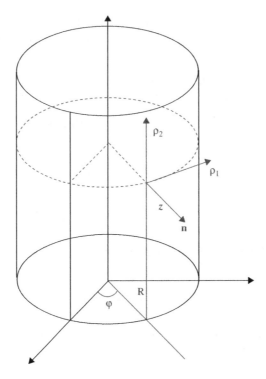

Figure 3.6 Cylinder and its local (surface) coordinates system $(\boldsymbol{\rho}_1, \boldsymbol{\rho}_2, \mathbf{n})$, $(\xi^1 \equiv \phi, \xi^2 \equiv z)$

The first tangent vector $\boldsymbol{\rho}_1$:

$$\boldsymbol{\rho}_1 = \frac{\partial \boldsymbol{\rho}}{\partial \xi_1} = \left\{ \begin{array}{c} -R\sin\varphi \\ R\cos\varphi \\ 0 \end{array} \right\} \qquad (3.71)$$

Second derivatives:

$$\boldsymbol{\rho}_{11} = \frac{\partial \boldsymbol{\rho}}{\partial \xi^1 \partial \xi^1} = \left\{ \begin{array}{c} -R\cos\varphi \\ -R\sin\varphi \\ 0 \end{array} \right\}; \qquad \boldsymbol{\rho}_{12} = \frac{\partial \boldsymbol{\rho}}{\partial \xi^1 \partial \xi^2} = \left\{ \begin{array}{c} 0 \\ 0 \\ 0 \end{array} \right\}. \qquad (3.72)$$

The second tangent vector $\boldsymbol{\rho}_2$:

$$\boldsymbol{\rho}_2 = \frac{\partial \boldsymbol{\rho}}{\partial \xi^2} = \frac{\partial \boldsymbol{\rho}}{\partial z} = \left\{ \begin{array}{c} 0 \\ 0 \\ 1 \end{array} \right\}. \qquad (3.73)$$

All second derivatives are zero $\boldsymbol{\rho}_{22} = 0$, $\boldsymbol{\rho}_{21} = 0$.

3.4.4.1 Covariant Components of the Metric Tensor

$$a_{11} = \rho_1 \cdot \rho_1 = R^2\sin^2\varphi + R^2\cos^2\varphi = R^2 \quad (3.74)$$

$$a_{12} = \rho_1 \cdot \rho_2 = 0 \quad (3.75)$$

From the last equation it follows that the coordinate lines are orthogonal.

$$a_{22} = \rho_2 \cdot \rho_2 = 1. \quad (3.76)$$

This component a_{22} informs of unit length of the ρ_2.
Covariant components are given by the matrix

$$[a_{ij}] = \begin{bmatrix} R^2 & 0 \\ 0 & 1 \end{bmatrix}. \quad (3.77)$$

The inverse matrix gives us, according to equation (3.31), contravariant components

$$[a^{ij}] = [a_{ij}]^{-1} = \begin{bmatrix} \frac{1}{R^2} & 0 \\ 0 & 1 \end{bmatrix}$$

The normal vector

$$\mathbf{N} = \rho_1 \times \rho_2 = \begin{bmatrix} \mathbf{i} & \mathbf{j} & \mathbf{k} \\ -R\sin\varphi & R\cos\varphi & 0 \\ 0 & 0 & 1 \end{bmatrix} = \mathbf{i}R\cos\varphi + \mathbf{j}R\sin\varphi \quad (3.78)$$

has the length

$$|\mathbf{N}|^2 = R^2\cos^2\varphi + R^2\sin^2\varphi = R^2. \quad (3.79)$$

A unit normal vector is derived as

$$\mathbf{n} = \frac{\mathbf{N}}{|\mathbf{N}|} = \mathbf{i}\cos\varphi + \mathbf{j}\sin\varphi. \quad (3.80)$$

3.4.4.2 Curvature Tensor Components

Following the definition in equation (3.46), the covariant components of the curvature tensor are calculated as

$$h_{11} = (\rho_{11} \cdot \mathbf{n}) = -R(\cos^2\varphi + \sin^2\varphi) = -R \quad (3.81)$$

other components are zero

$$h_{12} = \rho_{12} \cdot \mathbf{n} = 0; \quad h_{22} = (\rho_{22}) \cdot \mathbf{n} = 0. \quad (3.82)$$

Thus, the covariant components of the curvature tensor are given by the matrix

$$h_{ij} = \begin{bmatrix} -R & 0 \\ 0 & 0 \end{bmatrix} \qquad (3.83)$$

The generalized eigenvalue problem, equation (3.55), leads to the following determinant:

$$det|\mathbf{H} + k\mathbf{A}| = det(h_{ij} + k_{ij}) = det \begin{bmatrix} -R + kR^2 & 0 \\ 0 & k \end{bmatrix} = 0. \qquad (3.84)$$

$$k(kR^2 - R) = 0 \qquad (3.85)$$

Principal curvatures are roots of equation (3.85):

$$k_1 = 0; \quad k_2 = \frac{1}{R}. \qquad (3.86)$$

Thus, the cylindrical surface locally contains a straight line (with zero curvature $k_1 = 0$) and a line with constant radius of curvature R.

3.4.4.3 Christoffel Symbols

The Christoffel symbols for the cylindrical surface are computed via equation (3.63). First, we compute contravariant vectors according to equation (3.34)

$$\boldsymbol{\rho}^1 = a^{1j}\boldsymbol{\rho}_j = a^{11}\boldsymbol{\rho}_1 + a^{12}\boldsymbol{\rho}_2 = a^{11}\boldsymbol{\rho}_1 = \left\{ \begin{array}{c} -\dfrac{\sin\varphi}{R} \\ \dfrac{\cos\varphi}{R} \\ 0 \end{array} \right\}$$

$$\boldsymbol{\rho}^2 = a^{2j}\boldsymbol{\rho}_j = a^{21}\boldsymbol{\rho}_1 + a^{22}\boldsymbol{\rho}_2 = \left\{ \begin{array}{c} 0 \\ 0 \\ 1 \end{array} \right\}$$

and then we obtain the Christoffel symbols

$$\Gamma^k_{21} = \Gamma^k_{12} = (\boldsymbol{\rho}_{12} \cdot \boldsymbol{\rho}^k) = 0, \quad k = 1,2$$

and

$$\Gamma^1_{11} = \boldsymbol{\rho}_{11} \cdot \boldsymbol{\rho}^1 = \sin\varphi\cos\varphi - \sin\varphi\cos\varphi = 0 \qquad (3.87)$$

$$\Gamma^2_{11} = \boldsymbol{\rho}_{11} \cdot \boldsymbol{\rho}^2 = 0 \qquad (3.88)$$

$$\Gamma^k_{22} = 0, \quad k = 1,2. \qquad (3.89)$$

Differential Geometry

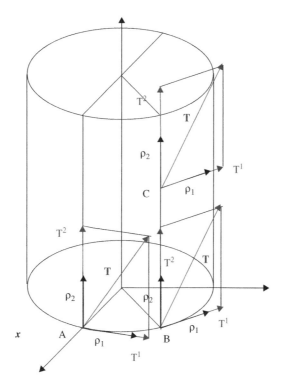

Figure 3.7 Geometrical interpretation of covariant derivative on the surface

3.4.4.4 Covariant Derivatives on a Cylindrical Surface: Geometrical Interpretation of the Parallel Transfer

As we computed before, all Christoffel symbols for the cylindrical surface are zero $\Gamma_{ij}^k = 0$ Thus, equation (3.66) is written as

$$\frac{d\mathbf{T}}{dt} = \left(\frac{\partial T^i}{\partial t} + \frac{\partial T^i}{\partial \xi^j}\dot{\xi}^j\right)\boldsymbol{\rho}_i + h_{11}T^1\dot{\xi}^1 \mathbf{n} \qquad (3.90)$$

here only non–zero components of h_{ij} from equation (3.66) are written.

Assume now that the vector $\mathbf{u}$ is constant on the surface. Thus, we can obtain for the surface vectors:

$$\boldsymbol{\rho}_\varphi: \quad \frac{\partial T^1}{\partial t} + \frac{\partial T^1}{\partial \varphi}\dot\varphi + \frac{\partial T^1}{\partial r}\dot r = 0 \qquad (3.91)$$

$$\boldsymbol{\rho}_z: \quad \frac{\partial T^2}{\partial t} + \frac{\partial T^2}{\partial \varphi}\dot\varphi + \frac{\partial T^2}{\partial z}\dot r = 0 \qquad (3.92)$$

or

$$T^1 = const \qquad (3.93)$$
$$T^2 = const. \qquad (3.94)$$

This gives us the geometrical interpretation of the covariant derivative as a parallel transfer.

Definition 3.4.2 *A vector* **T** *is transferred parallel to the surface (in the curvilinear surface coordinate system) if its components in this surface coordinate system remain constant. In this case covariant derivatives of those components are zero.*

The normal component of a vector on the surface in 3D space, however, is not zero and is computed as $h_{ij}T^1\xi^1$. In the current example it is computed as $h_{11}T^1\dot{\xi}^1 = -RT^1\dot{\varphi} \neq 0$, which expresses the part of the derivative outside the surface along the normal direction n.

4

Geometry and Kinematics for an Arbitrary Two Body Contact Problem

After preparing the necessary knowledge, both from classical mechanics and from differential geometry, we start now with the main goal of this textbook – to build computational contact mechanics algorithms. This process is based on the consideration of geometry and kinematics of two contacting bodies in a specially defined coordinate system. In a general 3D situation, this coordinate system is selected based on the geometrical properties of the contacting bodies leading to surface-to-surface, curve-to-surface, curve-to-curve (and so on) contact pairs. The fundamental basis in order to distinguish between those contact pairs is formed by the solution of the Closest Point Projection (CPP) procedure, namely by the existence and uniqueness of this solution. These general questions are far beyond the scope of the current *Introduction to Computational Contact Mechanics* and the interested reader is advised to read details in the book by Konyukhov and Schweizerhof (2012) *Computational Contact Mechanics – Geometrically Exact Theory for Arbitrary Shaped Bodies*. In the current book, we consider only a 2D analog of the "surface-to-surface" 3D contact pair leading to a "plane curve -to- plane curve" contact pair. The description will nevertheless follow the general strategy of the geometrically exact theory of contact interactions in such a way that after working with this course the reader can easily work with arbitrary 3D case algorithms for arbitrary geometrical situations.

Historically, through the development of the currently popular finite element codes (ABAQUS, ANSYS, LS-DYNA, etc.) the "master-slave" approach has been proved to be the most efficient approach in computational contact mechanics. Originally, the contact surface discretized by finite elements for one of the contacting bodies

---

Introduction to Computational Contact Mechanics: A Geometrical Approach, First Edition.
Alexander Konyukhov and Ridvan Izi.
© 2015 John Wiley & Sons, Ltd. Published 2015 by John Wiley & Sons, Ltd.
Companion Website: www.wiley.com/go/Konyukhov

was selected as a "master" surface, while a contacting surface from other body was considered as a cloud of "slave" nodes. The closest distance between the slave and the master found numerically as the solution of the Closest Point Projection (CPP) procedure has become a penetration – a measure of the normal contact interaction, see Remark 2.2.2 on the Signorini problem.

The terminology "master-slave" has become the standard in computational contact mechanics, though some variances such as "mortar part" for the "master body" and "non-mortar part" for the "slave" body are known in the literature.

In general, we define a *master* body on which the *observer* and, therefore, the local coordinate system will be placed. The *observed body* consists now of arbitrarily taken possible "slave" contact points and is called the *slave body*. The exact rule to identify the "slave" point depends on the further numerical method involved and it can be a "slave node", "slave integration point" and so on.

The main goal of the following description is to take advantage of the differential geometry of the contacting curves or surfaces in order to describe the geometry and kinematics of the contact including both the normal (penetration – non-penetration) and the tangential (sticking-sliding) interactions. We will concentrate in detail more on 2D contact, but will provide an overview, however of the general 3D description as well. Geometrically, contact for 2D kinematics is considered to be the interaction between two curves and for 3D kinematics is the interaction between two surfaces.

4.1 Local Coordinate System

Let us introduce two coordinate systems:

1. A global frame of reference as usual for the finite element discretization: XOY in case of 2D problems with coordinate axis X, Y, see Figure 4.1.
2. A local coordinate system attached to the master curve.

The first represents an inertial global frame of reference XOY (coordinates x, y) as usual for the finite element discretization and, the second coordinate system is a non-inertial local coordinate system attached to the contacting curve of the "master" body, see Figure 4.1. It is convenient to use two local coordinate systems:

1. The Frenet coordinate system with unit basis vectors τ and ν for the further analysis and derivations;
2. The coordinate system with basis vectors ρ_{ξ^1} of arbitrary length and n of the unit length for forthcoming finite element discretization.

Tangent vectors ρ_{ξ^1} and τ are parallel, while the unit vector n points outside of the master body, therefore, depending on the geometry (remember that the natural normal ν is pointing always into the center of curvature and, therefore, into the convex part of the body), the following equation is valid $n = \pm \nu$.

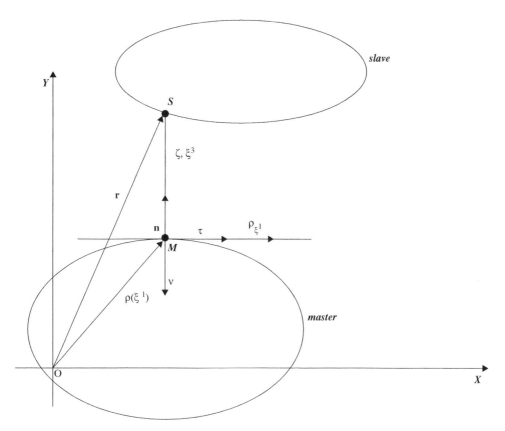

Figure 4.1 Local coordinate systems in 2D: (a) basis vectors ρ_ξ, $\mathbf{n}$ for arbitrary parametrization of the curve; (b) unit basis vectors τ and ν (or $\mathbf{n}$) for natural parametrization

The corresponding 2D coordinate system is constructed as follows:

1. Consider an arbitrarily chosen slave point **S** on the contacting boundary of the slave body;
2. Find the closest distance between the slave point **S** and the master curve. The closest point on the master surface denoted as point **C** represents the projection of the chosen slave point **S**. It is obtained as the solution of the Closest Point Projection procedure. The vectors for the master and slave points are denoted ρ and $\mathbf{r}$, respectively.
3. Introducing the variable ζ along the normal vector $\mathbf{n}$ pointing to the slave point, we can express the position vector of the slave point as

$$\mathbf{r}(\xi,\zeta) = \rho(\xi) + \zeta\mathbf{n}, \text{ or } \mathbf{r}(\xi^1,\xi^3) = \rho(\xi^1) + \xi^3\mathbf{n}. \tag{4.1}$$

In order to simplify further generalization the notations are adjusted for the 3D case as $\xi \equiv \xi^1$, $\zeta \equiv \xi^3$.

Equation (4.1) defines the mapping, or transformation of a pair of local coordinates ξ, ζ into a pair of Cartesian coordinates x, y, and therefore, defines a 2D local coordinate system. The uniqueness and existence of such a transformation in equation (4.1) depends on the uniqueness and existence of the corresponding CPP procedure.

The measures of the contact interaction are naturally defined in the local coordinate system equation (4.1) as follows:

- coordinate along normal **n** $\zeta \equiv \xi^3$ is the penetration – the measure of the normal interaction, see Remark 2.2.2 to the Signorini problem;
- coordinate $\xi \equiv \xi^1$ or arc-length coordinate $s(\xi)$ is responsible for the tangential interaction and is important for frictional problems.

The non-penetration condition is formulated as the following conditions

$$\zeta \equiv \xi^3 = \begin{cases} > 0, & \text{no contact} \\ = 0, & \text{slave point } S \text{ is on contact boundary} \\ < 0, & \text{contact with penetration.} \end{cases} \quad (4.2)$$

Remark 4.1.1 *For the non-frictional contact problems, non-penetration conditions in equation (4.2) represent only a single set of constraints that must be fulfilled. For frictional problems an additional constitutive relationship should be supplied, for example the Coulomb friction law.*

The coordinate system in equation (4.1) is defined at the projection point **C**. Thus, the first important auxiliary problem is to find the solution of the closest point projection (CPP) procedure.

4.2 Closest Point Projection (CPP) Procedure – Analysis

The Closest Point Projection (CPP) procedure is formulated directly as the extremal problem to find the shortest distance between the slave point **r** and the master curve $\rho(\xi)$:

$$F(\xi) := \frac{1}{2}\|\mathbf{r} - \boldsymbol{\rho}(\xi)\| = \frac{1}{2}(\mathbf{r} - \boldsymbol{\rho}(\xi)) \cdot (\mathbf{r} - \boldsymbol{\rho}(\xi)) \to min, \quad (4.3)$$

or formulated via the natural variable s

$$F(s) := \frac{1}{2}\|\mathbf{r} - \boldsymbol{\rho}(s)\| = (\mathbf{r} - \boldsymbol{\rho}(s)) \cdot (\mathbf{r} - \boldsymbol{\rho}(s)) \to min. \quad (4.4)$$

The necessary condition for minimum of the function F requires $\frac{dF}{d\xi} = 0$ for equation (4.3), or $\frac{dF}{ds} = 0$ for equation (4.4). Thus, we obtain

$$(\boldsymbol{\rho}(\xi) - \mathbf{r}) \cdot \frac{d\boldsymbol{\rho}(\xi)}{d\xi} = 0, \Rightarrow (\boldsymbol{\rho}(\xi) - \mathbf{r}) \cdot \boldsymbol{\rho}_\xi = 0 \quad (4.5)$$

or formulated via the arc-length

$$(\rho(s) - \mathbf{r}) \cdot \frac{d\rho(s)}{ds} = 0 \Rightarrow (\rho(s) - \mathbf{r}) \cdot \boldsymbol{\tau} = 0. \tag{4.6}$$

Definition of the unit tangent vector $\boldsymbol{\tau}$ in equation (3.3) is used to write down the final result in equation (4.6)

In general case of curvilinear geometry, equation (4.5) should be solved numerically. The form in equation (4.6) is used for further analysis of uniqueness and existence of the solution.

4.2.1 Existence and Uniqueness of CPP Procedure

It is easy to formulate the sufficient criteria for the minimum of the minimization problem in equation (4.4): at the point of minimum the second derivative of the minimized function F should be positive

$$\frac{d^2 F}{ds^2} = F'' > 0. \tag{4.7}$$

If condition (4.7) is fulfilled, then solution of the minimization problem (4.4) and, therefore, the solution of equation (4.6) exists and is unique. This allows us to define an *allowable projection domain* Ω in 2D as a set of Cartesian coordinates (x, y), for which representation in the coordinate system equation (4.1) is possible and the second derivative is positive for each point (ξ, ζ).

Remark 4.2.1 *Any point of the the allowable projection domain Ω is projected uniquely onto the given curve $\rho(\xi)$. The Newton iterative scheme, which is necessary to solve equation (4.5) numerically, is converging from any initial point on the curve $\xi_{(0)}$ reflecting the Cartesian point $x_{(0)}, y_{(0)}$, if the slave point S belongs to this domain.*

The second derivative of the function F in equation (4.4), see also equation (4.6),

$$F'' = \left(\frac{d\rho}{ds} \cdot \boldsymbol{\tau}\right) - (\mathbf{r} - \rho(s)) \cdot \frac{d\boldsymbol{\tau}}{ds} = 0 \tag{4.8}$$

is transformed in the local coordinate system using equation (4.1) and using the Frenet formula in equation (3.5) and equation (3.3) for the unit tangent vector $\boldsymbol{\tau}$. Finally

$$F'' = ((\boldsymbol{\tau} \cdot \boldsymbol{\tau}) - (\mathbf{r} - \rho(s)) \cdot k\boldsymbol{\nu} = 1 - \zeta k > 0. \tag{4.9}$$

Writing this criterion with the radius of curvature $R = \frac{1}{k}$ instead of the curvature k the following statement is obtained

$$1 - \frac{\zeta}{R} > 0. \tag{4.10}$$

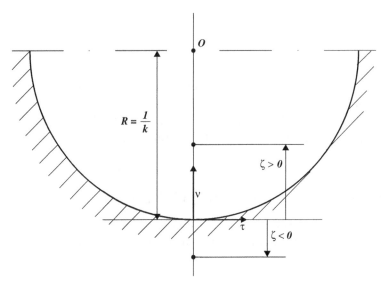

Figure 4.2 Allowable projection domains for the circular arch: (a) inner part $\{(x,y) \in \Omega \| \mathbf{r} = \boldsymbol{\rho} + \zeta \mathbf{n}, 0 < \zeta < R$, (b) outer part $\{(x,y) \in \Omega \| \mathbf{r} = \boldsymbol{\rho} + \zeta \mathbf{n}, \zeta < 0$

Since the normal vector $\boldsymbol{\nu}$ is always pointing to the center of curvature, we obtain from the condition (4.10) the following allowable projection domains, see the example Figure 4.2

- a finite projection domain for the convex part (above the curve in Figure 4.2):

$$\Omega(s, \zeta) := \left\{ \mathbf{r} = \boldsymbol{\rho}(s) + \zeta \boldsymbol{\nu}, \quad with \quad 0 < \zeta < \frac{1}{k} = R \right\}, \qquad (4.11)$$

- a semi-infinite projection domain for the concave part (below the curve in Figure 4.2):

$$\Omega(s, \zeta) := \{ \mathbf{r} = \boldsymbol{\rho}(s) + \zeta \boldsymbol{\nu}, \quad with \quad -\infty < \zeta < 0 \}. \qquad (4.12)$$

4.2.1.1 Allowable Projection Domain for the Circular Arch

The allowable projection domain Ω is constructed for any curve as a subset of the Cartesian space. In contact problems we have to distinguish which part is exactly filled with continuum (material). Thus, for the circular arch presented in Figure 4.2 all penetrating points $\zeta < 0$ are projected uniquely to the arch, however, one should be careful that the point can be represented in the given coordinate system in equation (4.1), see domains for the outer part of the quarter of a circle in Figure 4.3 and for the inner part in Figure 4.4 The allowable projection domain for an arch in Figure 4.4 includes the first quadrant of the Cartesian coordinate system with $x \geq 0, y \geq 0$. One can see, that it is not possible to project any point onto this arch from both the second

Geometry and Kinematics for an Arbitrary Two Body Contact Problem

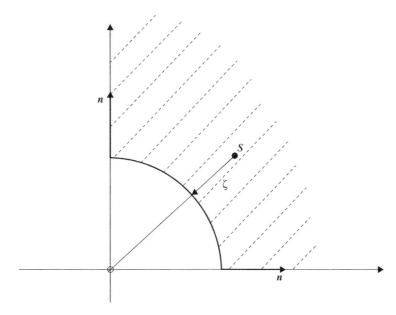

Figure 4.3 Projection domain for the outer part of the circle: $\{(x,y) \in \Omega \| \mathbf{r} = \boldsymbol{\rho} + \zeta\mathbf{n}, \zeta > 0\}$

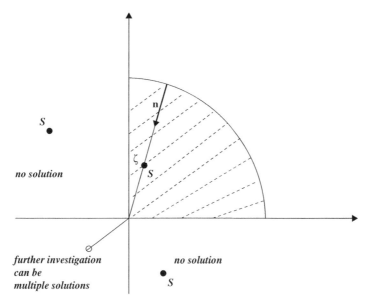

Figure 4.4 Projection domain for the inner part of the circle: $\{\mathbf{r} = (x,y) \in \Omega \| \mathbf{r} = \boldsymbol{\rho} + \zeta\mathbf{n}, 0 < \zeta < R\}$

quadrant $x < 0, y > 0$, and the fourth quadrant $x > 0, y < 0$. Any point from the third quadrant $x < 0, y < 0$ in Figure 4.4 does not satisfy the existence and uniqueness criteria and more advanced analysis is necessary – for the given arch this reflects the fact that, for the full circle (not only the given arch), it is possible to project a point twice; first, onto the given arch and, second, onto the opposite side of the circular arch – both solutions are found as solutions of the projection CPP procedure for the full circle on the radial line (of course, the shortest distance allows us to select only the one solution).

4.2.1.2 Irregular Cases

The existence and uniqueness of the CPP procedure requires the continuity of the derivatives of ρ, so-called, $C1$-continuity (the name $C1$ means *"the derivatives of the first order are continuous*, etc. for $C2$, $C3$). Discontinuities then require some further analysis. The discontinuities of the $C1$ type can be given either due to the real structure being discontinuous or due to the discretization of the boundary with linear elements. The discretization with linear elements causes $C1$-discontinuities between the element edges, see Figure 4.5. For both cases, the discontinuity makes a unique definition of the normal not possible, as shown in Figure 4.5. A further discontinuity of the $C2$ type might be regarding the curvature $\rho_{M_{\xi^1 \xi^2}}$, which is not that crucial since a normal vector is uniquely determined. In order to construct a numerical algorithm for a certain contact pair, first of all, it is identified that the closest distance between contacting bodies is a natural measure of the normal contact interaction. The procedure is introduced via the closest point projection (CPP) procedure, solution of which requires the differentiability of the function representing the parametrization of the surface of the contacting body. Analysis of the resolvability for the CPP procedure

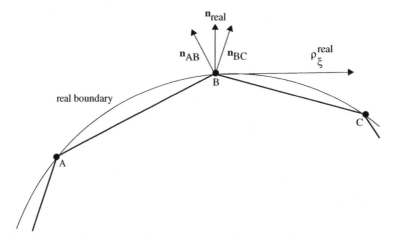

Figure 4.5 Linear FE representation of the boundary. Discontinuity of normals: n_{AB}, n_{BC}, and n_{real}

allows us then to classify all types of all possible contact pairs, see details in the book by Konyukhov and Schweizerhof (2012). The idea can easily shown for the 2D case in Figure 4.6. We consider a point belonging to the "slave" body and following the path $S_1\ S_2\ S_3\ S_4\ S_5$. The slave point is observed from the "master" side $\mathbf{ABCD}$, as we put an observer on this "master" line and providing, first, the Point-To-Curve CPP procedure only–we can say we observe the "slave" point in the following local coordinate system Point-To-Curve CPP procedure:

$$\mathbf{r}(s,\zeta) = \boldsymbol{\rho}(s) + \zeta\boldsymbol{\nu}. \tag{4.13}$$

Since a normal vector $\boldsymbol{\nu}$ has jumps (points $\mathbf{B}$ and $\mathbf{C}$), there are portions of the trajectory that can not be described in the local coordinate system inherited with the Point-To-Curve CPP procedure in equation (4.13). They are located in the non-allowable domain with regard to the projection onto the curve: any point S_2 laying in the non-allowable domain can not be described in the local coordinate system given by equation (4.13). The term *non allowable domain with regard to the projection onto surface resp. curve resp. point* is then used for a domain where any point can not be described in the local coordinate system corresponding to the projection onto surface resp. curve resp. point. However, in such a situation it is possible to create a continuous mapping of the path $S_1\ S_2\ S_3\ S_4\ S_5$ onto the curve by introducing a new projection operation in the non-allowable domain with regard to the projection onto the curve. This is a projection of the point into an angular point of curves (e.g., S_2 into B and S_4 into C in Figure 4.6), or CPP procedure onto the

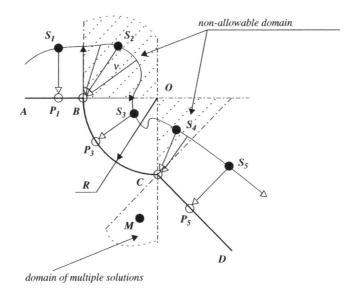

Figure 4.6 Violation of C^1-continuity leads to non-allowable domains with regard to the projection onto master curve $ABCD$. Both Point-To-Curve and Point-To-Point contact pairs are necessary to describe geometrically exact kinematics of contact with the master curve

corner points. This leads necessarily to Point-To-Point contact pairs. The projection domain for 3D case can be constructed for utmost C1-continuous surfaces. If the surfaces contain edges and vertexes then the CPP procedure should be generalized in order to include the projection onto edges and onto vertexes. The main idea for application for the contact is then straightforward–the CPP procedure corresponding to a certain geometrical feature gives a rise to a special, in general, a curvilinear 3D coordinate system. This coordinate system is attached to a geometrical feature and its convective coordinates are directly used for further definition of the contact measures. Thus, all contact pairs should be described in the corresponding local coordinate system. The requirement of the existence for the generalized CPP procedure leads to the transformation rule between types of contact pairs according to which the corresponding coordinate system is taken. Thus, all contact pairs can be uniquely described in most situations.

4.2.2 Numerical Solution of CPP Procedure in 2D

For the 2D case the CPP procedure is only depending on the surface coordinate ξ^1 and the Newton iterative scheme is written as

$$\Delta \xi^1_{(n+1)} = -\frac{F'}{F''}$$

$$\xi^1_{(n+1)} = \xi^1_{(n)} + \Delta \xi^1_{(n+1)} \tag{4.14}$$

where the derivatives are

$$F' = -(\mathbf{r} - \boldsymbol{\rho}(\xi^1)) \cdot \boldsymbol{\rho}_1$$
$$F'' = [(\boldsymbol{\rho}_1 \cdot \boldsymbol{\rho}_1) - (\mathbf{r} - \boldsymbol{\rho}(\xi^1)) \cdot \boldsymbol{\rho}_{11}] \tag{4.15}$$

4.2.3 Numerical Solution of CPP Procedure in 3D

For the inclusion of contact formulations within finite element algorithms, the closest point projection of an already fixed slave point on the master surface has to be considered and its distance to the slave point computed if penetration is identified. Due to discretization, numerical integration and procedure involved later, the check of penetration is carried out at integration points on the slave surface, see Chapter 7. The closest point projection procedure can be described by using the position vectors $\mathbf{r}$ and $\boldsymbol{\rho}$ of arbitrary points on the slave and master surface, respectively. The standard closest point projection procedure leads then to the following extremal problem in 3D

$$\|\mathbf{r} - \boldsymbol{\rho}(\xi^1, \xi^2)\| = (\mathbf{r} - \boldsymbol{\rho}(\xi^1, \xi^2)) \cdot (\mathbf{r} - \boldsymbol{\rho}(\xi^1, \xi^2)) \to min. \tag{4.16}$$

As is well known, the solution of this problem can be achieved by the application of the Newton procedure for the function $F(\xi^1, \xi^2) = \frac{1}{2}(\mathbf{r} - \boldsymbol{\rho}(\xi^1, \xi^2)) \cdot (\mathbf{r} - \boldsymbol{\rho}(\xi^1, \xi^2))$

where, within FE implementation, the initial estimate is the origin of the parent domain within the master surface element ($\xi^1 = 0;\ \xi^2 = 0$)

$$\begin{pmatrix} \Delta\xi^1 \\ \Delta\xi^2 \end{pmatrix}_{(n+1)} = -[F_i\ F_{ij}^{-1}]_{(n)} \tag{4.17}$$

$$\Rightarrow \begin{pmatrix} \xi^1 \\ \xi^2 \end{pmatrix}_{(n+1)} = \begin{pmatrix} \xi^1 \\ \xi^2 \end{pmatrix}_{(n)} + \begin{pmatrix} \Delta\xi^1 \\ \Delta\xi^2 \end{pmatrix}_{(n+1)}. \tag{4.18}$$

The first and second derivatives with respect to the surface coordinates ξ^i are given as

$$F_i = \begin{bmatrix} \dfrac{\partial F}{\partial \xi^1} \\ \dfrac{\partial F}{\partial \xi^2} \end{bmatrix} = -\begin{bmatrix} \rho_{\xi^1} \cdot (\mathbf{r}-\boldsymbol{\rho}) \\ \rho_{\xi^2} \cdot (\mathbf{r}-\boldsymbol{\rho}) \end{bmatrix} \tag{4.19}$$

$$F_{ij} = \begin{bmatrix} \dfrac{\partial^2 F}{\partial \xi^1 \partial \xi^1} & \dfrac{\partial^2 F}{\partial \xi^1 \partial \xi^2} \\ \dfrac{\partial^2 F}{\partial \xi^2 \partial \xi^1} & \dfrac{\partial^2 F}{\partial \xi^2 \partial \xi^2} \end{bmatrix} = \begin{bmatrix} a_{11} - \rho_{\xi^1\xi^1} \cdot (\mathbf{r}-\boldsymbol{\rho}) & a_{12} - \rho_{\xi^1\xi^2} \cdot (\mathbf{r}-\boldsymbol{\rho}) \\ a_{21} - \rho_{\xi^2\xi^1} \cdot (\mathbf{r}-\boldsymbol{\rho}) & a_{22} - \rho_{\xi^2\xi^2} \cdot (\mathbf{r}-\boldsymbol{\rho}) \end{bmatrix}$$

where the final result is gained according to the stop criteria for the iteration process with a given tolerance ϵ as

$$\left\| \begin{pmatrix} \Delta\xi^1 \\ \Delta\xi^2 \end{pmatrix} \right\| = \left\| \begin{pmatrix} \xi^1 \\ \xi^2 \end{pmatrix}_{(n+1)} - \begin{pmatrix} \xi^1 \\ \xi^2 \end{pmatrix}_{(n)} \right\| < \epsilon$$

or after a specific given amount of iterations. The obtained iterative solution ξ^1, ξ^2 in process in equations. (4.17) and (4.18) allows us to determine position of the projection point on the master surface as $\rho(\xi^1, \xi^2)$. The value of penetration can be calculated in the direction of normal vector $\mathbf{n}(\xi^1, \xi^2)$

$$\xi^3 = (\mathbf{r} - \boldsymbol{\rho}(\xi^1, \xi^2)) \cdot \mathbf{n}(\xi^1, \xi^2). \tag{4.20}$$

4.3 Contact Kinematics

Kinematics is considered in the 2D coordinate system: (coordinates ξ^i) defined locally on the closest point of the master surface to the slave surface point. The closest point on the master surface is, hereby, denoted as point C and represents a projection of a chosen slave point S. The position vector for the master and slave points are $\boldsymbol{\rho}_M$ and $\mathbf{r}_S$, respectively. Due to C representing the closest point projection of S, the normal $\mathbf{n}$ points to the slave point, thus, its position vector can be expressed in 2D as (see in Figure 4.7):

$$\mathbf{r}(\xi^1, \xi^3) = \boldsymbol{\rho}(\xi^1) + \xi^3 \mathbf{n}. \tag{4.21}$$

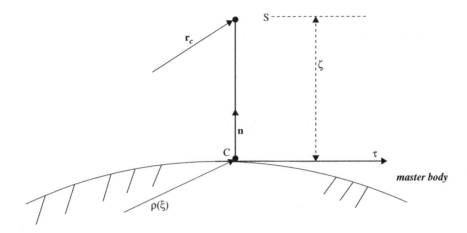

Figure 4.7 Contact kinematics for the 2D case

Considering the kinematics in 3D coordinate system (XYZ) in case of 3D problems with coordinates x, y, z, (see Figure 4.8), the following coordinate system is introduced

$$\mathbf{r}(\xi^1, \xi^2, \xi^3) = \boldsymbol{\rho}(\xi^1, \xi^2) + \xi^3 \mathbf{n},. \qquad (4.22)$$

where the master line in 2D and the master surface in 3D are parametrized with the Gaussian coordinates ξ^1 and (ξ^1, ξ^2), respectively.

Continuing now with 2D case, the tangent vector $\boldsymbol{\tau}$ at C can be constructed. A coordinate along the "master" line ξ^1 is a measure for the tangential interaction and is relevant for frictional problems. The coordinate $\zeta \equiv \xi^3$ represents the measure for normal interaction with the following notations

$$\xi^3 = \begin{cases} > 0, & \text{no contact} \\ = 0, & \text{slave point } S \text{ on contact boundary} \\ < 0, & \text{contact with penetration.} \end{cases}$$

All these considerations are based on the closest point C. This point is gained according to the closest point projection (CPP) procedure.

Regarding contact kinematics with moving and deforming contacting surfaces, all parameters are treated as time dependent, even for statical problems. For statical problems, time is treated as a load parameter. For the solution scheme of nonlinear equations, a linearization is required. Within the used geometrically exact description the increment vector is then treated as a velocity vector. In order to derive these velocities the motion process can be observed in the local coordinate system as a relative motion of the slave point S against the master surface. In order to take the

Geometry and Kinematics for an Arbitrary Two Body Contact Problem

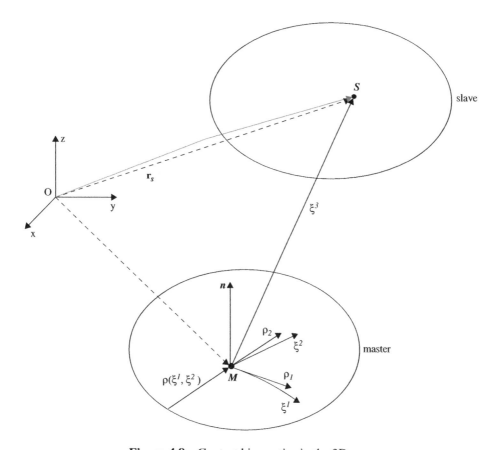

Figure 4.8 Contact kinematics in the 3D case

full time derivative of the position vector of the slave point equation (4.1) is used

$$\frac{d}{dt}\mathbf{r}(t, \xi^1, \xi^3) = \frac{d}{dt}\boldsymbol{\rho} + \frac{d}{dt}(\xi^3 \mathbf{n}) =$$

$$= \frac{\partial \boldsymbol{\rho}}{\partial t} + \boldsymbol{\rho}_{\xi^1}\dot{\xi}^1 + \dot{\xi}^3 \mathbf{n} + \xi^3 \frac{d\mathbf{n}}{dt}. \qquad (4.23)$$

Denoting the translational velocity of the master point M as $\mathbf{v}_M = \frac{\partial \boldsymbol{\rho}}{\partial t}$ and the absolute velocity of the slave point as $\mathbf{v}_S = \frac{d\mathbf{r}}{dt}$, equation (4.23) is transformed as

$$\mathbf{v}_S = \mathbf{v}_M + \boldsymbol{\rho}_{\xi^1}\dot{\xi}^1 + \dot{\xi}^3 \mathbf{n} + \xi^3 \frac{d\mathbf{n}}{dt}. \qquad (4.24)$$

The convective velocities $\dot{\xi}^1$ and the rate of penetration $\dot{\xi}^3$ as a projection on the local coordinate system of the master surface are obtained from equation (4.24) by evaluating the slave point velocity $\mathbf{v}_S$ at $\xi^3 = 0$ (ξ^3 is a small value enabling considerations on tangent plane) and taking the dot products to the basis vector $\boldsymbol{\rho}_{\xi^1}$ and $\mathbf{n}$ at the master point M. Thus, the projected convective velocities result in

$$\dot{\xi}^3 = (\mathbf{v}_S - \mathbf{v}_M) \cdot \mathbf{n} \tag{4.25}$$

$$\dot{\xi}^1 = \frac{(\mathbf{v}_S - \mathbf{v}_M) \cdot \boldsymbol{\rho}_{\xi^1}}{(\boldsymbol{\rho}_{\xi^1} \cdot \boldsymbol{\rho}_{\xi^1})} = m^{11}(\mathbf{v}_S - \mathbf{v}_M) \cdot \boldsymbol{\rho}_{\xi^1} \tag{4.26}$$

where the difference between the velocities $\mathbf{v}_S$ and $\mathbf{v}_M$ is a relative velocity $\mathbf{v}^{rel}$ of the slave point S in relation to M, or in other words, the velocity of point S as can be seen from point M. m^{11} represents the contravariant metric tensor.

Remark 4.3.1 (Differential geometry reminder – Metric tensor component in 2D) *The scalar multiplication of the covariant base vector $\boldsymbol{\rho}_{\xi^1}$ defines the covariant metric tensor of a cylindrical surface in 3D represented by a curve in 1D*

$$m_{11} = \boldsymbol{\rho}_{\xi^1} \cdot \boldsymbol{\rho}_{\xi^1}. \tag{4.27}$$

In 2D the contravariant metric tensor m^{11} can be obtained either by scalar multiplication of the contravariant base vector $\boldsymbol{\rho}^{\xi^1}$, or out of the relationship

$$m_{11}\, m^{11} = 1 \;\Rightarrow\; m^{11} = \frac{1}{m_{11}}.$$

The penetration rate $\dot{\xi}^3$ is computed exactly, even though computed not for tangent plane ($\xi^3 \neq 0$). Finally, the relative velocity vector on the tangent plane can be expressed as

$$\mathbf{v}_S - \mathbf{v}_M = \dot{\xi}^1 \boldsymbol{\rho}_{\xi^1} + \dot{\xi}^3 \mathbf{n}.$$

This approach of projected velocities on the tangent plane allows to consider the contact kinematics on the master surface only and to exploit the differential geometry of the surface during further considerations. This is a main feature of the used covariant description that leads to a simplification of the tangent matrix. It is important to note that only the contact problem is taken into account but not the motion and deformation of the two bodies system connected by means of the normal vector $\mathbf{n}$ at the distance ξ^3.

4.3.1 2D Contact Kinematics using Natural Coordinates s and ζ

An important part for formulation as well as for linearization of the weak form is a time derivative of the vector of a slave point $\mathbf{S}$

$$\frac{d\mathbf{r}_s}{dt} = \frac{\partial \boldsymbol{\rho}}{\partial t} + \dot{\xi}\boldsymbol{\rho}_\xi + \zeta\left(\frac{\partial \boldsymbol{\nu}}{\partial t} + \dot{\xi}\frac{\partial \boldsymbol{\nu}}{\partial \xi}\right) + \dot{\zeta}\boldsymbol{\nu}. \tag{4.28}$$

With $\mathbf{v}_s = \frac{d\mathbf{r}_s}{dt}$ as the absolute velocity of the slave point $\mathbf{S}$ resp. $\mathbf{v} = \frac{\partial \boldsymbol{\rho}}{\partial t}$ as the velocity of its projection on the master surface. The dot product with the normal vector $\boldsymbol{\nu}$ leads to the rate of the penetration

$$\dot{\zeta} = (\mathbf{v}_s - \mathbf{v}) \cdot \boldsymbol{\nu}. \tag{4.29}$$

Considering a value of the convective tangent velocity $\dot\xi$ on the tangent line, i.e. at $\zeta = 0$, we need the dot product of equation (4.28) with ρ_ξ leading to:

$$\dot\xi = \frac{(\mathbf{v}_s - \mathbf{v}) \cdot \rho_\xi}{(\rho_\xi \cdot \rho_\xi)}. \tag{4.30}$$

In the case of a length parametrization with $s = \xi$, equation (4.30) leads to the following convective velocity $\dot s$:

$$\dot s = (\mathbf{v}_s - \mathbf{v}) \cdot \boldsymbol\tau. \tag{4.31}$$

From the kinematic equation (4.28) we can obtain an equation for the variations by changing the time derivative operator into the variation operator δ. This equation is also considered on the tangent line, that is, at $\zeta = 0$:

$$\delta \mathbf{r}_s - \delta \boldsymbol\rho = \delta\xi \rho_\xi + \delta\zeta \boldsymbol\nu. \tag{4.32}$$

Equation (4.32) gives a variation of the displacement field for the expression of the virtual work of contact tractions on the contact surface.

4.3.2 Contact Kinematics in 3D Coordinate System

Some important formulas of contact interaction for kinematics in 3D are summarized in short as: the full time derivative of the vector $\mathbf{r}$

$$\frac{d\mathbf{r}(t, \xi^1, \xi^2, \xi^3)}{dt} = \frac{\partial \boldsymbol\rho}{\partial t} + \frac{\partial \boldsymbol\rho}{\partial \xi^i} \frac{d\xi^i}{dt} + \frac{\partial \mathbf{n}}{\partial t}\xi^3 + \frac{\partial \mathbf{n}}{\partial \xi^i}\frac{d\xi^i}{dt}\xi^3 + \mathbf{n}\frac{d\xi^3}{dt}$$

$$= \mathbf{v}_M + \rho_j \dot\xi^j + \frac{\partial \mathbf{n}}{\partial t}\xi^3 - h_i^j \rho_j \dot\xi^i \xi^3 + \mathbf{n}\dot\xi^3 \tag{4.33}$$

the relative velocity is

$$\mathbf{v}_S - \mathbf{v}_M = (\dot\xi^j - h_i^j \dot\xi^i \xi^3)\rho_j + \mathbf{n}\dot\xi^3 + \frac{\partial \mathbf{n}}{\partial t}\xi^3 \tag{4.34}$$

using the Weingarten formula in equation (3.60).
 The convective velocities are

$$\dot\xi^3 = (\mathbf{v}_S - \mathbf{v}_M) \cdot \mathbf{n} \tag{4.35}$$

$$\dot\xi^i = (\mathbf{v}_S - \mathbf{v}_M) \cdot \rho_k a^{ki}. \tag{4.36}$$

5

Abstract Form of Formulations in Computational Mechanics

Some new terminology and information from functional analysis, as well as from optimization theory, is necessary to describe both the formulation of contact problems and numerical methods in the most general form.

5.1 Operator Necessary for the Abstract Formulation

We introduce an operator $\mathbf{A}$ as the generalization of the standard function. If function $f(x)$ transforms a real number x into a real number $f(x)$, then an operator $\mathbf{A}$ transforms a member x of a space X into a member y of a space Y.

$$\mathbf{A}: X \to Y \tag{5.1}$$

Definition 5.1.1 *If $\mathbf{A}$ is an arbitrary (even nonlinear) operator then mapping in equation (5.1) is called the operator form equation. The simple form of this is similar to the linear algebra notation:*

$$y = \mathbf{A}x; \quad x \in X; \quad y \in Y \tag{5.2}$$

5.1.1 Examples of Operators in Mechanics

The examples of various operators can be written as follows:

1. Linear matrix:

$$\begin{aligned} y_i = a_{ij} x_j \quad & x = \{x_1, \ldots, x_n\} \in \mathbf{R}^n \text{ is a vector of } X \\ & y = \{y_1, \ldots, y_n\} \in \mathbf{R}^n \text{ is a vector of } Y \\ & a_{ij} \text{ is a component of the matrix } A \end{aligned} \tag{5.3}$$

Introduction to Computational Contact Mechanics: A Geometrical Approach, First Edition.
Alexander Konyukhov and Ridvan Izi.
© 2015 John Wiley & Sons, Ltd. Published 2015 by John Wiley & Sons, Ltd.
Companion Website: www.wiley.com/go/Konyukhov

2. Functional:

$$y = \int_0^l f(s)\, ds \quad f(s) \text{ is a member of a space } X\text{–integrable functions}$$

$$y \text{ is a real number, a member of a space } Y \qquad (5.4)$$

3. Functional – strain energy:

$$W = \int_{(v)} w(u_i) dV = \frac{1}{2}\int_{(v)} A_{ijkl}\varepsilon_{ij}\varepsilon_{kl} dV; \qquad (5.5)$$

where the linear strain tensor

$$\varepsilon_{ij} = \frac{1}{2}(u_{i,j} + u_{j,i}) \qquad (5.6)$$

u_i are components of a displacement vector – the space of differentiable functions, W is a real number – strain energy.

Definition 5.1.2 *Transformation $y = Ax$ is linear if*

(a) $A(\lambda x) = \lambda Ax$ where λ is a real number;
(b) $A(x_1 + x_2) = A(x_1) + A(x_2)$.

A is called a linear operator. The corresponding task, written only with the help of linear operators, is called a linear problem.

Remark 5.1.3 *In the Finite Element Method the following notation is used*

$$Ku = f. \qquad (5.7)$$

5.1.2 Examples of Various Problems

1. Linear elasticity problem is a linear problem.
2. Contact problem (Signorini) is a nonlinear problem, because;
   - unknown contact area S_c is not satisfying definition 5.1.2
   - contact stresses depend on the normal vector to the contact surface, the position of which depends on the deformed geometry.
3. Finite element formulation for the linear elasticity is the linear problem, because;

$$Ku = f; \quad K\text{–stiffness matrix}$$
$$u, f\text{–nodal vectors.} \qquad (5.8)$$

5.2 Abstract Form of the Iterative Method

For the solution of nonlinear problems in computational mechanics an iterative solution method to solve nonlinear problems written in operator $F(x) = 0$ includes the following steps:

1. Specification of a numerical method:
   Find a numerical operator A (fixed point operator) that corresponds to the nonlinear solution $F(x) = 0$:
   $$F(x) = 0 \iff x = Ax \tag{5.9}$$

2. The iterative solution is constructed as repetition of the numerical operator A to the initial approximation x_0. A single repetition is called iteration. It is required that
   $$F(x_n) \stackrel{n \to \infty}{\longrightarrow} 0 \tag{5.10}$$

Several questions arise:

1. Is the iterative solution converging, especially with regard to a selected initial value x_0?
2. Do we need a special choice for x_0?
3. Does this solution exist and is it unique?
$$F(x_*) = 0 \iff x_* = \lim_{n \to \infty} x_n \tag{5.11}$$
4. How fast is the iterative solution converging?

Since the given problems are nonlinear in computational contact mechanics as discussed before, an iterative solution method is required. Iterative solution methods are characterized by starting with a value x_0 for the unknown solution of the nonlinear problem and applying an operator A in order to gain x_1. The next step to use the same operator A for applying to x_1 in order to receive x_2 and going ahead with this iterative process as long as an *a priori* defined termination condition is not fulfilled. For iterative solution procedures in general the question arises; if these iterations are leading to the correct solution, how fast are they leading to this correct solution? Thus, the understanding of consistency, convergence and convergence rate are essential. While consistency expresses the fact that each result within an iterative solution scheme should be a solution of the given nonlinear problem, irrespectively from this convergence, it deals with the fact that independent from the starting value x_0 the iterative solution scheme has a unique solution. In case that both consistency and convergence are fulfilled and the iteration scheme reaches the unique solution of the given nonlinear problem with the speed of the convergence rate.

5.3 Fixed Point Theorem (Banach)

Consider an equation in the form
$$x = Ax \qquad (5.12)$$
where A is any nonlinear operator.

Example 5.3.1 *For the finite element method representing a stiffness matrix $Kx = f$ A is a matrix representing the transformation*
$$x = \left(E - \frac{K}{\lambda}\right)x + \frac{f}{\lambda} \qquad (5.13)$$
with any $\lambda \neq 0$ and with the unit matrix E.

Definition 5.3.2 $\|A\|$ *is a* norm of the operator A defined as,
$$\|A\| = \sup_{x \neq 0} \frac{\|Ax\|}{\|x\|} \qquad (5.14)$$

Definition 5.3.3 A *is an* operator-contractor, *if for any x, y the following expression*
$$\|Ax - Ay\| < \|A\|\|x - y\| \qquad (5.15)$$
is valid,
where $\|A\| < 1$.

Definition 5.3.4 *A fixed point of the operator A is called a solution of the equation $x^* = Ax^*$*

Theorem 5.3.5 (The fixed point theorem) *Any operator-contractor has a fixed point, computed via the iterative process*
$$x_{n+1} = Ax_n, \qquad (5.16)$$
$$\lim_{n \to \infty} x_n = x^* \qquad (5.17)$$

Proof.
$$\|x_{n+1} - x^*\| = \|Ax_n - Ax^*\| < \|A\|\|x_n - x^*\| < \|A\|^{n+1}\|x_0 - x^*\| < \varepsilon \qquad (5.18)$$
for all $\|A\| < 1 \Rightarrow x_n \to x^*$
with x_0 as an initial guess value

Definition 5.3.6 (Rate of convergence) *Consider a distance between the initial value x_0 and the solution x^**
$$q = \|x_0 - x^*\| \qquad (5.19)$$

A distance between $(n+1)$ approximation x_{n+1} and the solution x^* is defined as

$$I_{n+1} = \|x_{n+1} - x^*\| \tag{5.20}$$

It can be expressed via the initial distance q.
Introduce then a ratio:

$$r(q) = \frac{I_{n+1}}{I_n} \tag{5.21}$$

Degree of the polynom $r(q)$ is called the rate of convergence for the iterative process.

Theorem 5.3.7 *A fixed point iteration procedure for the operator-contractor has the constant rate of convergence.*

Proof.

$$r(q) = \frac{\|x_{n+1} - x^*\|}{\|x_n - x^*\|} \leq \frac{\|A\|\|x_n - x^*\|}{\|x_n - x^*\|} = \|A\| = const \tag{5.22}$$

Remark 5.3.8 *If A is differentiable and if $|A'(c)| < 1$ then A has a fixed point.*

$$|Ax - Ay| = |A'(c)(x - y)| \leq |A'(c)|\|x - y| \tag{5.23}$$

for arbitrary points $x, y \in D$ and a suitable $c \in D$ (mid-point rule).

5.4 Newton Iterative Solution Method

Consider nonlinear, in general case, equation

$$F(x) = 0 \tag{5.24}$$

x represents, for example, a vector or a point
 Let x^* be a solution
 Consider a value $x^* + \Delta x$ close to the solution

$$F(x^* + \Delta x) \approx 0 \tag{5.25}$$

Assuming that F is differentiable, expand into the Taylor series and construct an iterative process as follows

$$F(x_n) + F'(x_n)\Delta x = 0 \tag{5.26}$$

$$x_{n+1} - x_n = \Delta x$$

$$F(x_n) + F'(x_n)(x_{n+1} - x_n) = 0 \tag{5.27}$$

Assume that $(F'(x_n))^{-1}$ exists

$$x_{n+1} = x_n - (F'(x_n))^{-1} F(x_n) \tag{5.28}$$

This iterative method is called the *Newton method*. The method is converging if the initial point x_0 is close to the solution point.

Remark 5.4.1 *Equation (5.28) is in general valid in the operator form.*

Remark 5.4.2 *Using finite element method (FEM) terminology*

$R = F(x_n)$ *is called a residual vector. It represents the equilibrium equation for nonlinear problems.*
$K = F'(x_n)$ *is called a tangent matrix. It represents the stiffness of the structure.*

5.4.1 Geometrical Interpretation of the Newton Iterative Method

The geometrical interpretation of equation (5.28) in the case of a function of one variable is an iterative approaching to the solution x^* using the intersection of the tangent line started at the point x_0 with the axis OX, see Figure (5.1).

One can see if the function $F(x)$ possesses certain properties so that then the convergence is very fast and the solution $F(x) = 0$ is unique. This is observed for the so-called convex function.

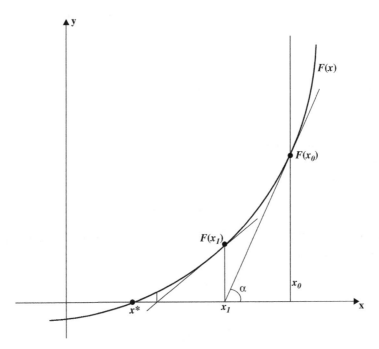

Figure 5.1 Geometrical interpretation of the Newton Iterative Solution Method: $x_1 = x_0 - (F'(x_0))^{-1} F(x_0) = x_0 - \dfrac{F(x_0)}{\tan \alpha}$, while $\tan \alpha = \dfrac{F(x_0)}{x_0 - x_1}$

Definition 5.4.3 (The convex Function $F(x)$) *The function $F(x)$ is convex if, for any A, B laying along the graph $F(x)$ and $0 < \lambda < 1$, a point $C = \lambda A + (1 - \lambda)B$ lays inside the graph, see the representation in Figure 5.2. (Short mathematical notation is $\forall A, B \in \Omega, C = \lambda A + (1 - \lambda)B \in \Omega, \forall \lambda \in [0, 1]$.)*

Remark 5.4.4 (Examples of the convex function of one variable.) *If $F''(x) > 0$ then $F(x)$ is strictly convex.*

Remark 5.4.5 (Convergence of the Newton method.) *If $F(x)$ is strictly convex in $x \in D$ then equation $F'(x) = 0$ has a unique solution in D and the Newton method is converging from any $x_0 \in D$.*

Theorem 5.4.6 *The convergence rate for the Newton method. If the Newton method converges and*

$$\|(F'(x))^{-1}\| < c_1 \tag{5.29}$$

$$\|F''(x)\| < c_2 \tag{5.30}$$

then the rate of convergence is quadratic.

Proof. Consider the initial distance

$$I_0 = \|x_0 - x^*\| = q \tag{5.31}$$

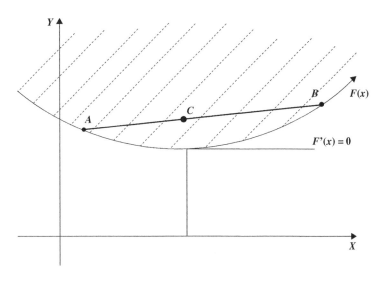

Figure 5.2 Convex function and its convex domain

Expand the following values into Taylor series:
$$F(x_0) = F(x_0 + x^* - x^*) = F(x^*) + F'(x^*)(x_0 - x^*)$$
$$+ \frac{1}{2}F''(x^*)(x_0 - x^*)(x_0 - x^*) + o_1(\|x_0 - x\|^2) \tag{5.32}$$
$$F'(x^*) = F'(x_0 + x^* - x_0) = F'(x_0) + F''(x^* - x_0) + o_2(\|x^* - x_0\|) \tag{5.33}$$

The next distance is computed via the Newton iteration as
$$I_1 = \|x_1 - x^*\| = \|x_0 - (F'(x_0))^{-1}F(x_0) - x^*\|$$
$$= \|x_0 - (F'(x_0))^{-1}[F'(x^*)(x_0 - x^*)$$
$$+ \frac{1}{2}F''(x^*)(x_0 - x^*)(x_0 - x^*) + o_1(\|x_0 - x^*\|^2)] - x^*\|$$
$$= \|x_0 - x^* - F'(x_0)^{-1}[\{F'(x_0) + F''(x_0)(x^* - x_0) + o_2\}(x_0 - x^*)$$
$$+ \frac{1}{2}F''(x^*)(x_0 - x^*)(x_0 - x^*) + o_1]\|$$
$$= \|x_0 - x^* - (x_0 - x^*) - F'(x_0)^{-1}[-F''(x_0)(x^* - x_0)(x^* - x_0)$$
$$+ \frac{1}{2}(F''(x^*)(x^* - x_0)(x^* - x_0) + o(\|x^* - x_0\|^2)]\|$$
$$= \|F'(x_0)^{-1}(\frac{1}{2}F''(x^*) - F''(x_0))(x^* - x_0)(x^* - x_0) + o(\|x^* - x_0\|^2)\|$$
$$\leq c_1 c_2 q^2 = C_1 q^2 \tag{5.34}$$

where we started with
$$x_0 - (F'(x_0))^{-1}F(x_0) = x_1 \tag{5.35}$$

Continuing recursively one can easily prove that:
$$I_1 = \|x_1 - x^*\| \leq C_1 q^2$$
$$I_2 = \|x_2 - x^*\| \leq C_2 q^4$$
$$\ldots$$
$$I_n = \|x_n - x^*\| \leq C_n q^{2n} \tag{5.36}$$

because if x_0 is close to the solution and the Newton method converges then
$$I_2 = \|x_2 - x^*\| \leq C_1 \|x^* - x_1\|^2$$
$$\leq \tilde{C}_1 C_1 \|x^* - x_0\|^4 = C_2 \|x^* - x_0\|^4 = C_2 q^4 \tag{5.37}$$

Now it is easy to prove that the rate of convergence is two
$$r(q) = \frac{I_{n+1}}{I_n} = \frac{C_{n+1} q^{2(n+1)}}{C_n q^{2n}} = \frac{C_{n+1}}{C_n} q^2. \tag{5.38}$$

5.5 Abstract Form for Contact Formulations

The Signorini problem discussed in Section 2.2 can be specified in the abstract form for the contact formulations. The minimization problem using more general operator form is given as

$$F(u) = \frac{1}{2} Au \cdot u - f(u) \rightarrow \min \tag{5.39}$$

subjected to the inequality condition for the contact boundary Γ_c

$$Bu - g \leq 0. \tag{5.40}$$

Remark 5.5.1 *This operator form is valid not only for the contact formulation, written using differential operations such as in continuum mechanics, but also for the contact formulation, formulated after the finite element discretization.*

5.5.1 Lagrange Multiplier Method in Operator Form

As already shown in Section 1.3.1 the Lagrange multiplier method is constructed as a sum of the functional that should be minimized and the constraint conditions multiplied with the Lagrange multiplier λ

$$L(u, \lambda) = \frac{1}{2} Au \cdot u - f(u) + (Bu - g) \cdot \lambda \rightarrow \min. \tag{5.41}$$

Definition 5.5.2 (Saddle point.) *A pair of u, λ satisfying the minimization problem including the Karush–Kuhn–Tucker conditions is called a saddle point of the Lagrange functional $L(u, \lambda)$. This task of constrained minimization is called the saddle point problem.*

The Karush–Kuhn–Tucker conditions corresponding to the constraint equation (5.40) are written in the following form:

$$Bu - g \leq 0 \tag{5.42}$$

$$\lambda \geq 0$$

$$\lambda \cdot (Bu - g) \quad 0 \tag{5.43}$$

The necessary requirement of the minimum in equation (5.41) is that the partial derivatives are zero:

$$\begin{cases} \dfrac{\partial L}{\partial u} = Au - f' + B\lambda = 0 \\ \dfrac{\partial L}{\partial \lambda} = Bu - g = 0. \end{cases} \tag{5.44}$$

The second equation expresses the exact fulfillment of the non-penetration condition $p = Bu - g = 0$.

An operator form in a more compact notation for equation (5.44) is formulated as follows

$$\begin{bmatrix} A & B \\ B & 0 \end{bmatrix} \begin{pmatrix} u \\ \lambda \end{pmatrix} = \begin{pmatrix} f' \\ g \end{pmatrix} \qquad (5.45)$$

$$\underbrace{\phantom{\begin{bmatrix} A & B \\ B & 0 \end{bmatrix}}}_{\mathbf{M}} \quad \underbrace{\phantom{\begin{pmatrix} u \\ \lambda \end{pmatrix}}}_{\mathbf{x}} \quad \underbrace{\phantom{\begin{pmatrix} f' \\ g \end{pmatrix}}}_{\mathbf{f}} \qquad (5.46)$$

Remark 5.5.3 (Example: The statement of the Signorini problem)

$$\begin{cases} \int \sigma_{ij} \delta \epsilon_{ij} d\Omega - \delta W^{\Gamma_\sigma} - \delta W^f + \lambda_k \delta u_{(n)_k} = 0 \\ \delta \lambda_i (u_{(n)} - g)_i = 0. \end{cases} \qquad (5.47)$$

The following questions arise:

- Does the solution of the saddle point problem exist?
- Is the solution unique?

These answers are given by the following theorem:

Theorem 5.5.4 (Babuska–Brezzi condition) *If the following conditions are satisfied*

1. $A = A^T$ *is symmetric (as e.g. the stiffness matrix* $\mathbf{K}$*),*
2. *A is positive, that is*
   $Au \cdot v \leq c \|u\| \|v\|$
   is fulfilled for any arbitrary u, v and $c > 0$,
3. *B is bounded above, that is*
   $\|Bu\| \leq c\|u\|$
   is fulfilled for any arbitrary u and $c > 0$,
4. *B is bounded below, that is*
   $\|Bu\| \geq \gamma \|u\|$
   is fulfilled for any arbitrary u and $\gamma > 0$.
   (γ is called inf-sup constant).

then the solution of the saddle point problem

$$\begin{bmatrix} A & B \\ B & 0 \end{bmatrix} \mathbf{x} = \mathbf{f} \qquad (5.48)$$

exists and is unique.

5.5.2 Penalty Method in Operator Form

Consider the definition of the penalty function

Definition 5.5.5 *The penalty function $P(p)$ is a function of the constraint $p = Bu - g$ and it includes a constant penalty parameter ε. The penalty function satisfies the following conditions:*

1. *$P(0) = 0$ the function is zero if constraints are satisfied $p = 0$,*
2. *$P(p) \geq 0$ the function is positive for any p,*
3. *$\frac{dP}{du}$ exists with ϵ as a constant penalty parameter.*

The standard representation of the penalty function is

$$P(u) = \frac{1}{2}\varepsilon(Bu - g)(Bu - g). \tag{5.49}$$

The Penalty method in the operator form is formulated as follows

$$F_p = \frac{1}{2}Au \cdot u - f(u) + \varepsilon\frac{1}{2}(Bu - g)(Bu - g) \to \min. \tag{5.50}$$

Due to the functional depending on only one variable u, the necessary requirement for the minimizing of equation (5.50) is

$$\frac{dF_p}{du} = Au - f'(u) + \varepsilon(Bu - g) = 0 \tag{5.51}$$

$$\Rightarrow (A + \epsilon B)u = f'(u) + \epsilon g. \tag{5.52}$$

As for the Lagrange method before, the Signorini statement would look like

$$\int \sigma_{ij}\delta\varepsilon_{ij}d\Omega - \delta W^{\Gamma_\sigma} - \delta W^f + \varepsilon u_k \delta u_k = 0. \tag{5.53}$$

It can be seen that for $\varepsilon \to \infty$ the exact solution for the contact problem can be gained.

Theorem 5.5.6 *Under the assumptions in Definition 5.5.5 for the penalty function $P(u)$ there exists a sequence u_ε of such solutions that is converging to the solution of the original problem*

$$\lim_{\varepsilon \to \infty} F_p(u) = \min F(u). \tag{5.54}$$

6

Weak Formulation and Consistent Linearization

The contact kinematics developed earlier in a special coordinate system in a covariant form the basis for a variational form or a weak form of the contact problem and its further linearization. In this chapter, the weak form for frictional contact problem is derived in a covariant form in both natural and convective coordinates. The weak form is then linearized using covariant derivatives in the local coordinate system. Special attention is given to linearization independent of discretization in a covariant for both the penalty and Lagrange multipliers methods. Application to the normal following forces case is shown as a particular case of the "inverted" contact case. In addition, the Nitsche method is formulated in a covariant form as an alternative method fulfilling the non-penetration condition exactly.

6.1 Weak Formulation in the Local Coordinate System

Let $\mathbf{R}_A$ and $\mathbf{R}_B$ be the contact traction vectors acting on the elementary contact curve, see Figure 6.1. Consider two infinitesimal curves ds_A and ds_B representing boundaries of 2D bodies coming into contact. The resulting virtual work δW^c of the contact forces on the considered virtual displacements $\delta \mathbf{u}_A, \delta \mathbf{u}_B$ is obtained by the following surface integrals

$$\delta W^c = \int_{s_A} \mathbf{R}_A \cdot \delta \mathbf{u}_A ds_A + \int_{s_B} \mathbf{R}_B \cdot \delta \mathbf{u}_B ds_B. \qquad (6.1)$$

This work of contact forces must be added to the global virtual work of the internal and external forces. During the contact, equilibrium condition should be fulfilled at each point of contact:

$$\mathbf{R}_A ds_A + \mathbf{R}_B ds_B = 0. \qquad (6.2)$$

Introduction to Computational Contact Mechanics: A Geometrical Approach, First Edition.
Alexander Konyukhov and Ridvan Izi.
© 2015 John Wiley & Sons, Ltd. Published 2015 by John Wiley & Sons, Ltd.
Companion Website: www.wiley.com/go/Konyukhov

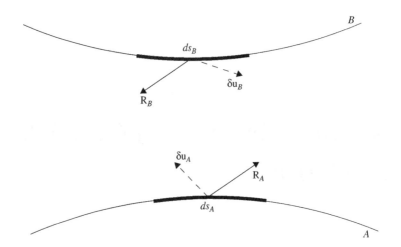

Figure 6.1 Equilibrium during contact. Infinitesimally small contact lines: ds_A; ds_B

Thus, the integral in equation (6.1) representing the weak form for the contact can be reformulated either as an integral over the contact line s_A or along the contact line s_B. Considering the integral over s_B gives

$$\delta W^c = \int_s \mathbf{R}_B \cdot (\delta \mathbf{u}_B - \delta \mathbf{u}_A) ds_B. \tag{6.3}$$

Now consider the weak form (6.3) in the local coordinate system for the contact description defined in Chapter 4 choosing the contact surface s_A as the master body and s_B as the slave body. The contact integral can be written as

$$\delta W^c = \int_s \mathbf{R} \cdot (\delta \mathbf{r}_s - \delta \boldsymbol{\rho}) ds \tag{6.4}$$

replacing $\delta \mathbf{u}_B$ by $\delta \mathbf{r}$ and $\delta \mathbf{u}_A$ by $\delta \boldsymbol{\rho}$ as the variation of the slave and master point, respectively. We selected integration over the slave part. The traction vector $\mathbf{R}$ in the local coordinate system can be split into a normal and tangential part

$$\mathbf{R} = \mathbf{N} + \mathbf{T} = N\mathbf{n} + T^1 \boldsymbol{\rho}_1. \tag{6.5}$$

Here, T^1 represents the contravariant component of the tangential contact surface traction vector $\mathbf{T}$. This vector in general is not described within a unit coordinate system and the absolute value of the tangential part is given as $\mid \mathbf{T} \mid = T \mid \boldsymbol{\rho}_1 \mid$ where T has the physical dimension $Dim(T) = \frac{Dim(\mathbf{T})}{Dim(\boldsymbol{\rho}_1)}$. Regarding $\mathbf{N}$, this is not the case, for $\mathbf{N}$ goes along with a unit vector $\mathbf{n}$.

An expression for the variation of the relative displacement vectors $\delta \mathbf{r} - \delta \boldsymbol{\rho}$ is derived following the analogy to the velocity in equation (4.24) on the master surface ($\xi^3 = 0$) as

$$\delta \mathbf{r} - \delta \boldsymbol{\rho} = \boldsymbol{\rho}_1 \delta \xi^1 + \mathbf{n} \delta \xi^3. \tag{6.6}$$

Then the contact integral in equation (6.4) can be written in the following form

$$\delta W^c = \int_s [\underbrace{N\delta\xi^3}_{\text{normal part}} + \underbrace{T^1(\boldsymbol{\rho}_1 \cdot \boldsymbol{\rho}_1)\delta\xi^1}_{\text{frictional part}}]ds. \tag{6.7}$$

In the case of non-frictional problems the tangential part disappears ($T^1 = 0$) and the integral is reduced to

$$\delta W_N^c = \int_s N\delta\xi^3 ds. \tag{6.8}$$

6.2 Regularization with Penalty Method

The contact traction components N and T^1 in equation (6.7) are, in general, additional unknowns of the contact integral. If they are treated as independent variables, the Lagrangian multiplier method is used, as we discussed for the example given in Chapter 1. If the traction is treated by the regularization scheme then a dependence with respect to the corresponding measure of contact interaction is supplied. For the non-frictional case in equation (6.8) this leads with the penalty regularization discussed before

$$N = \varepsilon_N H(-\xi^3)\xi^3 \text{ with } H(-\xi^3) = \begin{cases} 0 \\ 1 \end{cases}, \text{if } \begin{array}{l} \xi^3 > 0 \\ \xi^3 \leq 0 \end{array} \tag{6.9}$$

where ε_N is a penalty parameter and $H(-\xi^3)$ is the Heaviside function expressing that a normal force is existing only if the value of penetration ξ^3 is non-positive. The contact integral in equation (6.8) is transformed using the regularization scheme in equation (6.9) as

$$\delta W_N^c = \int_s \varepsilon_N H(-\xi^3)\xi^3 \delta\xi^3 ds. \tag{6.10}$$

6.3 Consistent Linearization

Non linear problems such as contact problems are solved numerically via iterative solution methods. The most common here is the Newton method. This method requires the derivative of the weak form δW_N^c. The procedure of taking derivatives is called linearization. The result with the following finite element discretization is a tangent matrix for the finite element implementation. The full time derivative in the form of covariant derivatives in the local coordinate system is used to derive the tangent matrix in the geometrically exact theory. Namely, usage of the covariant derivatives preserves the frame indifference.

6.3.1 Linearization of Normal Part

The linearization of the regularized contact functional in equation (6.10) follows then to

$$D_v(\delta W_N^c) = D_v\left(\int_s \varepsilon_N H(-\xi^3)\xi^3\delta\xi^3\,ds\right) =$$

$$= \underbrace{\int_s \varepsilon_N H(-\xi^3)\dot{\xi}^3\delta\xi^3\,ds}_{D_{\dot{\xi}^3}}$$

$$+ \underbrace{\int_s \varepsilon_N H(-\xi^3)\xi^3\delta\dot{\xi}^3\,ds}_{D_{\delta\dot{\xi}^3}}$$

$$+ \underbrace{\int_s \varepsilon_N H(-\xi^3)\xi^3\delta\xi^3\,(\dot{ds})}_{D_{(\dot{ds})}}. \tag{6.11}$$

We consider the linearization terms $D_{\dot{\xi}^3}$, $D_{\delta\dot{\xi}^3}$ and $D_{(\dot{ds})}$ separately.

The linearization of the integral $D_{(\dot{ds})}$ vanishes due to the fact that the integral is defined on the slave body approximated by the given numerical integration scheme (a cloud of integration points), but derivatives (leading to the linearization) are taken for the master surface. Thus, each integration point η_i has a projected fixed master point ξ_i^1 according to the CPP procedure and the linearization term $D_{(\dot{ds})}$ gives us

$$D_{(\dot{ds})} = \int_s \varepsilon_N H(-\xi^3)\xi^3\delta\xi^3(\dot{ds}) = \sum \varepsilon_N H(-\xi^3)\xi^3(\xi^1{}_i)\delta\xi^3(\xi^1{}_i)\frac{d}{dt}(J(\eta_i)w_i) = 0.$$

This is also emphasized by the structure of global numerical solution, where all integration points η_i on the slave segment are fixed during the iterative solution scheme.

The linearization term $D_{\dot{\xi}^3}$, together with the linearization of ξ^3, is evolving to

$$D_{\dot{\xi}^3} = \int_s \varepsilon_N H(-\xi^3)(\delta \mathbf{r} - \delta\boldsymbol{\rho})\cdot\mathbf{n}(\mathbf{v}_S - \mathbf{v})\cdot\mathbf{n}\,ds$$

considering equation (4.25) and the analogy between variation and velocity vectors, see equation (4.32).

The following tensor transformation is necessary in order to continue further.

Definition 6.3.1 *The dyadic product of two vectors* **a** *and* **b** *is written by components as*

$$[\mathbf{a}\otimes\mathbf{b}] = \begin{bmatrix} a_1b_1 & a_1b_2 & a_1b_3 \\ a_2b_1 & a_2b_2 & a_2b_3 \\ a_3b_1 & a_3b_2 & a_3b_3 \end{bmatrix}.$$

Lemma 6.3.2 *Tensor transformation of scalar products*

$$(\mathbf{a}\cdot\mathbf{b})(\mathbf{c}\cdot\mathbf{d}) = \mathbf{a}\cdot(\mathbf{b}\otimes\mathbf{c})\mathbf{d} = \mathbf{b}\cdot(\mathbf{a}\otimes\mathbf{c})\mathbf{d} = \mathbf{b}\cdot(\mathbf{a}\otimes\mathbf{d})\mathbf{c}. \tag{6.12}$$

Weak Formulation and Consistent Linearization

this gives

$$D_{\dot{\xi}^3} = \int_s \varepsilon_N H(-\xi^3)(\delta \mathbf{r} - \delta \boldsymbol{\rho}) \cdot (\mathbf{n} \otimes \mathbf{n})(\mathbf{v}_S - \mathbf{v}) \, ds.$$

The second term $D_{\delta\dot{\xi}^3}$ in equation (6.11) covers the linearization of $\delta\dot{\xi}^3$. The result will be obtained assuming a natural parametrization of the corresponding boundary curve $\xi^3 = \zeta$, $\xi^1 = s$. Using the natural parametrization we have to take into account that the natural normal is always pointed into the center of curvature, therefore

$$\mathbf{n} = \pm \boldsymbol{\nu}. \tag{6.13}$$

Taking the time derivative of the variational analog of equation (4.29) we obtain

$$\frac{d}{dt}\delta\zeta = \frac{d}{dt}[(\delta\mathbf{r}_s - \delta\boldsymbol{\rho}) \cdot \mathbf{n}] = \frac{d}{dt}[(\delta\mathbf{r}_s - \delta\boldsymbol{\rho}) \cdot (\pm\boldsymbol{\nu})] =$$

$$= \pm \left(\frac{\partial(\delta\mathbf{r}_s - \delta\boldsymbol{\rho})}{\partial s} \cdot \boldsymbol{\nu}\dot{s} + (\delta\mathbf{r}_s - \delta\boldsymbol{\rho}) \cdot \frac{\partial\boldsymbol{\nu}}{\partial t} + (\delta\mathbf{r}_s - \delta\boldsymbol{\rho}) \cdot \frac{\partial\boldsymbol{\nu}}{\partial s}\dot{s} \right). \tag{6.14}$$

The first term can be rewritten, taking into account (4.31) for a convective velocity in the case of a natural parametrization, as follows:

$$\delta\underbrace{\frac{\partial(\mathbf{r}_s - \boldsymbol{\rho})}{\partial s}}_{-\boldsymbol{\tau}} \cdot \boldsymbol{\nu}\dot{s} = -(\delta\boldsymbol{\tau} \cdot \boldsymbol{\nu})\underbrace{((\mathbf{v}_s - \mathbf{v}) \cdot \boldsymbol{\tau})}_{\dot{s}} = -\delta\boldsymbol{\tau} \cdot (\boldsymbol{\nu} \otimes \boldsymbol{\tau})(\mathbf{v}_s - \mathbf{v}). \tag{6.15}$$

In order to rewrite the second term, we have to take first a partial time derivative of the orthogonality condition:

$$\boldsymbol{\tau} \cdot \boldsymbol{\nu} = 0 \Rightarrow \frac{\partial(\boldsymbol{\tau} \cdot \boldsymbol{\nu})}{\partial t} = \frac{\partial^2\boldsymbol{\rho}}{\partial s \partial t} \cdot \boldsymbol{\nu} + \frac{\partial\boldsymbol{\nu}}{\partial t} \cdot \boldsymbol{\tau} = 0, \tag{6.16}$$

leading to the expression

$$\frac{\partial\boldsymbol{\nu}}{\partial t} \cdot \boldsymbol{\tau} = -\frac{\partial\mathbf{v}}{\partial s} \cdot \boldsymbol{\nu}. \tag{6.17}$$

From the other side, using the unity condition of the vector $\boldsymbol{\nu}$, we can express the time derivative using Lemma 3.1.1 in terms of the tangent vector $\boldsymbol{\tau}$ first, as

$$\frac{\partial\boldsymbol{\nu}}{\partial t} = a\boldsymbol{\tau} \tag{6.18}$$

and then obtaining a via the scalar product as

$$a = \left(\boldsymbol{\tau} \cdot \frac{\partial\boldsymbol{\nu}}{\partial t} \right) \boldsymbol{\tau} \tag{6.19}$$

finally equation (6.18) is written as

$$\frac{\partial\boldsymbol{\nu}}{\partial t} = \left(\boldsymbol{\tau} \cdot \frac{\partial\boldsymbol{\nu}}{\partial t} \right) \boldsymbol{\tau}. \tag{6.20}$$

The last equation (6.20) is used to transform equation (6.17) as

$$\frac{\partial \boldsymbol{\nu}}{\partial t} = -\left(\frac{\partial \mathbf{v}}{\partial s} \cdot \boldsymbol{\nu}\right) \boldsymbol{\tau}. \qquad (6.21)$$

Equation (6.21) allows to transform the second term in (6.14) as follows

$$(\delta \mathbf{r}_s - \delta \boldsymbol{\rho}) \cdot \frac{\partial \boldsymbol{\nu}}{\partial t} = -(\delta \mathbf{r}_s - \delta \boldsymbol{\rho}) \cdot \left(\frac{\partial \mathbf{v}}{\partial s} \cdot \boldsymbol{\nu}\right) \boldsymbol{\tau} = \qquad (6.22)$$

introducing a tensor product $\boldsymbol{\tau} \otimes \boldsymbol{\nu}$ in order to transform a dot product

$$= -(\delta \mathbf{r}_s - \delta \boldsymbol{\rho}) \cdot (\boldsymbol{\tau} \otimes \boldsymbol{\nu}) \frac{\partial \mathbf{v}}{\partial s} = -(\delta \mathbf{r}_s - \delta \boldsymbol{\rho}) \cdot (\boldsymbol{\tau} \otimes \boldsymbol{\nu}) \frac{\partial \boldsymbol{\tau}}{\partial t}. \qquad (6.23)$$

The last term in (6.23) is obtained reversing the order of differentiation as

$$\frac{\partial \mathbf{v}}{\partial s} = \frac{\partial}{\partial s}\frac{\partial \boldsymbol{\rho}}{\partial t} = \frac{\partial}{\partial t}\frac{\partial \boldsymbol{\rho}}{\partial s} = \frac{\partial \boldsymbol{\tau}}{\partial t}. \qquad (6.24)$$

The third term in (6.14) is reorganized into a tensor form with a second Serret–Frenet formula and with equation (4.31) for the convective velocity $\dot{s}$:

$$(\delta \mathbf{r}_s - \delta \boldsymbol{\rho}) \cdot \frac{\partial \boldsymbol{\nu}}{\partial s} \dot{s} = -(\delta \mathbf{r}_s - \delta \boldsymbol{\rho}) \cdot \kappa \boldsymbol{\tau} \otimes \boldsymbol{\tau} (\mathbf{v}_s - \mathbf{v}). \qquad (6.25)$$

Therefore, combining (6.15), (6.22) and (6.25), we obtain a final formula for the linearization of $\delta \zeta$:

$$\frac{d}{dt}\delta\zeta = \pm\left\{-\left(\delta\boldsymbol{\tau}\cdot\boldsymbol{\nu}\otimes\boldsymbol{\tau}(\mathbf{v}_s-\mathbf{v})+(\delta\mathbf{r}_s-\delta\boldsymbol{\rho})\cdot\boldsymbol{\tau}\otimes\boldsymbol{\nu}\frac{\partial\boldsymbol{\tau}}{\partial t}\right)\right.$$
$$\left.-(\delta\mathbf{r}_s-\delta\boldsymbol{\rho})\cdot\kappa\boldsymbol{\tau}\otimes\boldsymbol{\tau}(\mathbf{v}_s-\mathbf{v})\right\}. \qquad (6.26)$$

Going back to the contact kinematics description with the normal $\mathbf{n}$ (see equation (6.13)), the expression in coordinates ξ^1, ξ^3 in equation (6.26) is written as

$$\delta\dot{\xi}^3 = -\delta\boldsymbol{\tau}\cdot\mathbf{n}\otimes\boldsymbol{\tau}(\mathbf{v}_s-\mathbf{v}) - (\delta\mathbf{r}-\delta\boldsymbol{\rho})\cdot\boldsymbol{\tau}\otimes\mathbf{n}\frac{\partial\mathbf{v}}{\partial s} \mp k(\delta\mathbf{r}-\delta\boldsymbol{\rho})\cdot\boldsymbol{\tau}\otimes\boldsymbol{\tau}(\mathbf{v}_s-\mathbf{v}). \qquad (6.27)$$

Difficulties with a double sign before the curvature part (minus for concave and plus for convex) can be automatically resolved using reduction of the full 3D description to the 2D case. In this case a single component h^{11} of the curvature tensor is taken into account. Summarizing these linearization terms and inserting back to equation (6.11) the linearization of the weak form δW_N^c results in

$$D_v(\delta W_N^c) = D_{\dot{\xi}^3} + D_{\delta\dot{\xi}^3}$$
$$= \int_s (\varepsilon_N \ H(-\xi^3) \ \{(\delta\mathbf{r}-\delta\boldsymbol{\rho})\cdot(\mathbf{n}\otimes\mathbf{n})(\mathbf{v}_S-\mathbf{v}) \qquad (6.28)$$

$$-\xi^3 \left[\delta\boldsymbol{\tau} \cdot (\mathbf{n} \otimes \boldsymbol{\tau})(\mathbf{v}_S - \mathbf{v}) + (\delta\mathbf{r} - \delta\boldsymbol{\rho}) \cdot (\boldsymbol{\tau} \otimes \mathbf{n}) \frac{\partial \mathbf{v}}{\partial s} \right] \quad (6.29)$$

$$\pm \xi^3 k (\delta\mathbf{r} - \delta\boldsymbol{\rho}) \cdot (\boldsymbol{\tau} \otimes \boldsymbol{\tau})(\mathbf{v}_S - \mathbf{v})) ds. \quad (6.30)$$

The Frenet formulas are formulated for the unit tangent $\boldsymbol{\tau}$ and the unit normal $\boldsymbol{\nu}$ pointed necessarily in the direction of the center of curvature. In an arbitrary contact situation, a normal vector can be pointed arbitrarily, thus it can be written as $\mathbf{n} = \pm \boldsymbol{\nu}$ depending on the situation. Here the complete linearization is represented in subdivisions of a "main" part equation (6.28), a "rotational" part equation (6.29) and a "curvature" part equation (6.30). The main part represents a constitutive law for the normal interaction force N and, thus, is always present. The rotational part consists of two terms being transposed with each other, as will be seen after the discretization. The rotational characteristic can be observed clearly by the variation of the tangent unit vector on the master $\delta\boldsymbol{\tau}$ as $\delta\boldsymbol{\tau} \cdot \boldsymbol{\tau} = 0$ being perpendicular to the tangential vector any time and therefore describing a rotation. The variation $\delta\boldsymbol{\tau}$ and, thus, the rotational part vanishes when the master being rigid and fixed or, *a priori*, only translated. The curvature part is active for a curved master segment and vanishes for discretization of the master segment with linear elements ($k = 0$). Both the rotational and the curvature part can be neglected for small displacement problems. From another side, if contact force N is an independent variable, then only the rotational and curvature parts are presented (no main part at all!). This case appears for the Lagrange multipliers method or in the particular case of following normal forces. In the case of following forces, the force is just given as an external force, which always remains normal in the local coordinate system $\mathbf{N} = N\mathbf{n}$.

Remark 6.3.3 *Within a finite element computation the main contribution is given by the "main" part –the contact kinematics (non-penetration) will be observed correctly even if only the main part is implemented.*

Remark 6.3.4 *The finite element discretization of the linearized weak form equations (6.28–6.30) results in the tangent matrix for the implementation into a nonlinear finite element code.*

6.4 Application to Lagrange Multipliers and to Following Forces

Though the rotational part for the linearized contact integral using the Penalty method plays a minor auxiliary role in order to keep only full consistency, there are situations in which *the rotational part plays a major role*. They are:

- Lagrange multipliers method.
- Application of the result to following forces.

6.4.1 Linearization for the Lagrange Multipliers Method

If the Lagrange multipliers method is involved then the normal traction N in equation (6.8) is the independent variable and is included in linearization additional to $\dot N$ instead of the *main part*. The weak form remains unchanged as in equation (6.8) and its linearization in equations (6.28), (6.28), (6.30) becomes

$$D_v(\delta W_N^c)|_{Lagr.} =$$

$$= \int_s \dot N H(-\xi^3)(\delta \mathbf{r} - \delta \boldsymbol{\rho}) \cdot \mathbf{n}\, ds \qquad (6.31)$$

$$- \int_s N H(-\xi^3) \left[\delta\boldsymbol{\tau} \cdot (\mathbf{n} \otimes \boldsymbol{\tau})(\mathbf{v}_S - \mathbf{v})\, ds, + (\delta \mathbf{r} - \delta \boldsymbol{\rho}) \cdot (\boldsymbol{\tau} \otimes \mathbf{n}) \frac{\partial \mathbf{v}}{\partial s} \right]$$

$$(6.32)$$

$$\pm \int_s N H(-\xi^3) k (\delta \mathbf{r} - \delta \boldsymbol{\rho}) \cdot (\boldsymbol{\tau} \otimes \boldsymbol{\tau})(\mathbf{v}_S - \mathbf{v}))\, ds. \qquad (6.33)$$

Remark 6.4.1 *Within a finite element implementation for the simplest linear problem, the discretization of the main part in equation (6.31) plays a major role together with the discretization of the KKT condition in the form $\int_s N\dot\xi^3 = 0$. For an example of this implementation within the Node-To-Node (NTN) contact approach, see Section 14.1 in Chapter 14, Part II.*

6.4.2 Linearization for Following Forces: Normal Force or Pressure

Consideration of the local coordinate system in the covariant form allows us to model the situation with a normal force or pressure that remains always normal during the deformation. Such a case is known as a normal following force or pressure. The covariant form allows us to obtain both the residual and the tangent matrix in a straightforward fashion (compare, e.g., with the continuum mechanics cases in the monograph by Bonet and Wood (1999). Let us consider a normal force/pressure given in the local coordinate system as

$$\mathbf{N} = N\mathbf{n} \qquad (6.34)$$

then the weak form in equation (6.8) remains unchanged as

$$\delta W_N^c = \int_s N \delta \xi^3 ds, \qquad (6.35)$$

but its linearization contains then only the rotational part

$$D_v(\delta W_N^c)|_{foll.f.} =$$

$$- \int_s N \left[\delta \boldsymbol{\tau} \cdot (\mathbf{n} \otimes \boldsymbol{\tau})(\mathbf{v}_S - \mathbf{v}) + (\delta \mathbf{r} - \delta \boldsymbol{\rho}) \cdot (\boldsymbol{\tau} \otimes \mathbf{n}) \frac{\partial \mathbf{v}}{\partial s} \right] \qquad (6.36)$$

$$\pm N k (\delta \mathbf{r} - \delta \boldsymbol{\rho}) \cdot (\boldsymbol{\tau} \otimes \boldsymbol{\tau})(\mathbf{v}_S - \mathbf{v}))\, ds. \qquad (6.37)$$

Remark 6.4.2 *Implementation of the residual in equation (6.35) and its linearization in equations (6.36–6.37) allows us to check from one side the correctness of the implementation for the rotational part and from another side to model the application of the single normal force or the normal pressure. An example of the simple implementation of the single normal force and cross-verification of the rotational part is given in the example of Following forces based on the Node-To-Segment (NTS) contact approach in Section 16.3 in Part II. The single normal following force can be generally implemented based on the segment-wise approach similar to STAS, see Section 17.3.1, the constant following distributed in Lagrangian metrics (constant pressure in the local coordinate system) is given in the example of Pressure based on the STAS approach, see Section 17.3.2 and, finally, the example of the inflating beam widely used, for example for hydro-forming simulations, is given in the example of Inflating, see Section 17.3.3 in Chapter 17, Part II.*

6.5 Linearization of the Convective Variation $\delta\xi$

The linearization of convective variations $\delta\xi^i$ in a 3D formulation is the most complicated part of the process, see details in the monograph by Konyukhov and Schweizerhof (2012).

The main points of the linearization process are:

1. The convective variations are defined on the tangent plane of the spatial coordinate system via the consideration of the slave point velocity as $\dot{\xi}^j = a^{ij}(\mathbf{v}_s - \mathbf{v}) \cdot \boldsymbol{\rho}_i$.
2. During the linearization of $\delta\xi^i$ the derivative of the metric tensor is obtained as derivative of the spatial metric tensor considering its value on the tangent plane.

The reduction to the specific 2D plane geometry leads to the following equation:

$$\frac{d}{dt}(\delta\xi) =$$

$$= -\frac{(\delta\mathbf{r}_s - \delta\boldsymbol{\rho}) \cdot \boldsymbol{\rho}_\xi \otimes \boldsymbol{\rho}_\xi \ \mathbf{v}_j + \delta\boldsymbol{\rho}_\xi \cdot \boldsymbol{\rho}_\xi \otimes \boldsymbol{\rho}_\xi \ (\mathbf{v}_s - \mathbf{v})}{(\boldsymbol{\rho}_\xi \cdot \boldsymbol{\rho}_\xi)^2} \quad (6.38a)$$

$$+ \frac{(\boldsymbol{\rho}_{\xi\xi} \cdot \boldsymbol{\nu})}{(\boldsymbol{\rho}_\xi \cdot \boldsymbol{\rho}_\xi)^2}(\delta\mathbf{r}_s - \delta\boldsymbol{\rho}) \cdot (\boldsymbol{\rho}_\xi \otimes \boldsymbol{\nu} + \boldsymbol{\nu} \otimes \boldsymbol{\rho}_\xi)(\mathbf{v}_s - \mathbf{v}) + \quad (6.38b)$$

$$+ \frac{h_{11}}{a_{11}}\dot{\zeta}\delta\xi - \frac{\boldsymbol{\rho}_{\xi\xi} \cdot \boldsymbol{\rho}_\xi}{(\boldsymbol{\rho}_\xi \cdot \boldsymbol{\rho}_\xi)}\dot{\xi}\delta\xi. \quad (6.38c)$$

6.6 Nitsche Method

Among other methods to satisfy the constraint equation (5.40) exactly for the minimization problem (5.39), the Nitsche method is known as a method that is independent from the penalty parameter or stabilization parameter, ε_N. Here, we overview only briefly the idea of the derivation for this method, the more detailed derivation is given in the book Konyukhov and Schweizerhof (2012).

This method can be expressed as a version of the Lagrange multiplier method presented in Section 5.5.1. For the penalty method, we added the non-penetration constraint ($\xi^3 = 0$) to the functional. However, for the Nitsche method, the condition $\boldsymbol{\sigma}_M \cdot \boldsymbol{n}_M + \boldsymbol{\sigma}_S \cdot \boldsymbol{n}_S = \boldsymbol{0}$ is also added as an additional constraint to the functional using the Lagrange multipliers method. This constraint is valid on the contact boundary of contacting bodies. Thus, the functional follows as

$$N(u, \lambda, \mu) = \frac{1}{2} \int_\Omega \boldsymbol{\sigma} : \boldsymbol{\varepsilon}\, d\Omega - \int_\Omega \boldsymbol{f} \cdot \boldsymbol{u}\, d\Omega + \int_{\Gamma_C} \boldsymbol{\lambda} \cdot (\boldsymbol{r}_S - \boldsymbol{\rho}_M)\, d\Gamma$$
$$+ \int_{\Gamma_{C_M}} \boldsymbol{\mu}_M \cdot (\boldsymbol{\sigma}_M \cdot \boldsymbol{n}_M)\, d\Gamma + \int_{\Gamma_{C_S}} \boldsymbol{\mu}_S \cdot (\boldsymbol{\sigma}_S \cdot \boldsymbol{n}_S)\, d\Gamma \to \min$$

where $\boldsymbol{\mu}_M$ and $\boldsymbol{\mu}_S$ represent a second set of Lagrange multipliers on the master and slave surface, respectively, going along not with the contact stress as for λ within the Lagrange multiplier method, but with the contact kinematics $\boldsymbol{\mu}_M = -\boldsymbol{\rho}_M$ and $\boldsymbol{\mu}_S = -\boldsymbol{r}_S$. Considering also only non-frictional problems the contact contributions to the functional are expressed as

$$\ldots \int_{\Gamma_C} \underbrace{\lambda(\boldsymbol{r}_S - \boldsymbol{\rho}_M) \cdot \boldsymbol{n}}_{\xi^3}\, d\Gamma + \int_{\Gamma_C} (\boldsymbol{r}_S - \boldsymbol{\rho}_M) \cdot \underbrace{\underbrace{[(\boldsymbol{\sigma}_M \cdot \boldsymbol{n}_M) \cdot \boldsymbol{n}_M]}_{\sigma_{n_M} n_M}}_{\sigma_{n_M} \xi^3}\, d\Gamma.$$

A specific formulation for the Nitsche method, called the Gauss point-wise substituted formulation, is achieved by enforcing the part of the non-penetration condition ($\xi^3 = 0$) by the penalty type of formulation as

$$N(u) = \frac{1}{2} \int_\Omega \boldsymbol{\sigma} : \boldsymbol{\varepsilon}\, d\Omega - \int_\Omega \boldsymbol{f} \cdot \boldsymbol{u}\, d\Omega$$
$$+ \int_{\Gamma_C} \frac{1}{2} \varepsilon_N (\xi^3)^2\, d\Gamma + \int_{\Gamma_C} \sigma_{n_M} \xi^3\, d\Gamma \to \min. \tag{6.39}$$

Hereby, ε_N is denoted as the stabilization parameter that has no influence on the result of the computation compared to the Penalty method. As one can see from equation (6.39), the functional depends only on displacement values and not on further unknowns as Lagrange multipliers, even though they originated as a mixed formulation. Therefore, the weak form of the contact contribution in the normal direction for the Nitsche formulation again uses the Heaviside function

$$\delta W_N^c = \int_s \varepsilon_N H(-\xi^3) \xi^3 \delta\xi^3\, ds + \int_s H(-\xi^3) \left(\sigma_N \delta\xi^3 + \xi^3 \delta\sigma_N \right) ds, \tag{6.40}$$

where σ_N is the projection of the stress vector of the master body in direction of the normal $\boldsymbol{n}$.

6.6.1 Example: Independence of the Stabilization Parameter

We will illustrate the independence of the stabilization parameter ε_N and the exact fulfillment of the constraint condition (non-penetration) using the example from Chapter 1, presented in Figure 1.1. The Nitsche part in equation (6.40) should be added to the variation of the energy $\delta\Pi(u)$ in equation (1.4).

$$\delta\Pi(u) + \delta W_N^c = 0$$

$$\Rightarrow \delta\left(\frac{1}{2}cu^2 - mgu\right) + \varepsilon_N \xi^3 \delta\xi^3 + (\sigma_N \, \delta\xi^3 + \xi^3 \, \delta\sigma_N) = 0 \quad (6.41)$$

in order to get the final equation. Now we have to pay attention to the correct understanding of all parameters in the last equation.

The penetration and its variation is taken directly from equation (1.5):

$$\xi^3 = p(u) = l + u - H$$
$$\delta\xi^3 = \delta u \quad (6.42)$$

The contact stress σ_N can be computed as the inner stress computed in due course as a derivative of the potential energy $\Pi(u)$ in equation (1.4) times the normal vector of the contact boundary, using the Cauchy stress theorem in equation (2.4):

$$\sigma_N = \frac{d\Pi}{du} \cdot (-1) = -cu + mg$$
$$\delta\sigma_N = -c\delta u \quad (6.43)$$

Now inserting everything into equation (6.41) we obtain:

$$(cu - mg)\delta u + \varepsilon_N(l + u - H)\delta u - (cu - mg)\delta u - (l + u - H)c\delta u = 0$$
$$(\varepsilon_N - c)(l + u - H)\delta u = 0 \quad (6.44)$$

And now amazingly we obtain the exact fulfillment of the non-penetration condition

$$p(u) = (l + u - H) = 0 \quad (6.45)$$

for all values of ε_N, except the case $\varepsilon_N = c$.

Such an effect is shown in the numerical example in Chapter 15 in Part II, considering the implementation in Section 7.3.

Remark 6.6.1 *Though, the Nitsche method shows the advantages of both the Lagrange multipliers method, such as the exact fulfillment of the non-penetration condition, and the Penalty method, such as dependence on only the primal variable u, the necessity to implement the whole structural finite element (because of σ_N) positions the Nitsche method among the very seldomly used methods in practical FE codes.*

7
Finite Element Discretization

This chapter contains the most important information for further finite element implementation and studies various approaches on how to discretize the weak form in order to obtain the residual and how to discretize the linearized equations in order to obtain tangent matrices. Both the weak form in equation (6.7) (see Chapter 6) and the linearized equations (6.28–6.30) (see Section 6.3) should be discretized in order to obtain the corresponding system of equations for further iterative solution within the finite element method. Different discretization approaches for contact problems are available, among the most used and established are the "Node-To-Node" (NTN), "Segment-To-Analytical-Surface" (STAS), "Node-To-Surface" (NTS) and "Surface-To-Surface" or "Segment-To-Segment" (STS) Mortar approaches. Each of these discretization techniques can be chosen separately from the structural discretization, thus leading to the contact finite element, which can be independent from the finite elements specified for the structural part. The discretization approaches described in this chapter concentrate on the pure contact part of finite element computation. The difference between various discretization techniques is based on the method for checking penetration against the discretized parts. Using the master and slave body concept (see Chapter 4), the NTN approach checks the contact between a node on the slave and a node on the master body. The STAS approach checks the contact between the discretized master surface and an analytically described surface of a slave body, and the NTS approach checks the contact between a slave node against a discretized master surface. The STS Mortar approach checks the contact between the approximated slave surface and the approximated master surface. In this chapter, we intentionally concentrate only on the type of discretization and consider only non-frictional problems.

Introduction to Computational Contact Mechanics: A Geometrical Approach, First Edition.
Alexander Konyukhov and Ridvan Izi.
© 2015 John Wiley & Sons, Ltd. Published 2015 by John Wiley & Sons, Ltd.
Companion Website: www.wiley.com/go/Konyukhov

7.1 Computation of the Contact Integral for Various Contact Approaches

Various discretization approaches are distinguished by the contact type of penetration check as well as by the numerical integration of corresponding integrals appearing in both the weak form and consistent linearization. We consider the following approaches:

1. Node-To-Node (NTN)
2. Node-To-Segment (NTS)
3. Segment-To-Analytical-Segment (STAS)
4. Segment-To-Segment (STS) Mortar.

7.1.1 Numerical Integration for the Node-To-Node (NTN)

In case of the Node-To-Node approach, the contact is checked between single nodes of the master and slave body. Thus, for equation (6.10) the integral vanishes

$$\delta W_N^c = \int_s \varepsilon_N H(-\xi^3) \xi^3 \delta \xi^3 ds = \varepsilon_N H(-\xi^3) \xi^3 \delta \xi^3. \tag{7.1}$$

7.1.2 Numerical Integration for the Node-To-Segment (NTS)

In case of the Node-To-Segment approach, the contact is checked between a single slave node and the master segment. In equation (6.10), the integral is computed at the active node, that is, at its projection

$$\delta W_N^c = \int_s \varepsilon_N H(-\xi^3) \xi^3 \delta \xi^3 ds = \varepsilon_N H(-\xi^3) \xi^3 \delta \xi^3 |_{\xi_C}. \tag{7.2}$$

A node is determined as being active at the current segment if the following conditions are satisfied:

- The node is projected on the master projection point ξ_C and lies inside the master segment, that is $|\xi_C| \leq 1$ in many cases of approximations.
- The penetration ξ^3 computed at this node is negative (this fact is taken into account by the Heaviside function $H(-\xi^3)$).

7.1.3 Numerical Integration for the Segment-To-Analytical Segment (STAS)

In case of the Segment-To-Analytical Segment approach, the contact is checked between each integration point of the master segment and the analytically described segment. In equation (6.10), the integral is computed numerically as a sum over active integration points ξ_i positioned at the approximated segment (deformable body)

$$\delta W_N^c = \int_s \varepsilon_N H(-\xi^3)\xi^3 \delta\xi^3 ds = \sum_{i=1}^N \varepsilon_N H(-\xi^3)\xi^3 \delta\xi^3 \sqrt{(\rho_\xi \cdot \rho_\xi)}|_{\xi_i} w_i. \quad (7.3)$$

An active integration point is determined as the point at which penetration is negative – this fact is taken into account by the Heaviside function $H(-\xi^3)$.

7.1.4 Numerical Integration for the Segment-To-Segment (STS)

In case of the Segment-To-Segment approach, the contact is checked between each integration point of the slave segment and the master segment. In equation (6.10) the integral is computed numerically as a sum over active integration points η_i positioned at the slave segment – all parameters, in due course, are computed at the points projected on the master segment

$$\delta W_N^c = \int_s \varepsilon_N H(-\xi^3)\xi^3(\xi^1)\, \delta\xi^3(\xi^1)\, ds(\eta)$$

$$= \sum_{i=1}^N \varepsilon_N H(-\xi^3)\xi^3(\xi_i^1)\, \delta\xi^3(\xi_i^1)\, \sqrt{(\mathbf{r}_\eta \cdot \mathbf{r}_\eta)}|_{\eta_i} w_i. \quad (7.4)$$

An integration point η_i is determined as being active at the current segment if the following conditions are satisfied:

- An integration point η_i is projected on the master projection point ξ_i and lies inside the master segment, that is, $|\xi_i| \leq 1$ in many cases of approximations.
- The penetration ξ^3 computed at this point is negative (this fact is taken into account by the Heaviside function $H(-\xi^3)$).

Remark 7.1.1 *Contact area is approximated by contact points, penetrating into the current master segment and defined on the slave body.*

Remark 7.1.2 *All contact approaches can be written in a standardized form using the approximation matrix $[A]$ discretizing the two surfaces.*

Each approach will be explained in the following in more detail, with the concentration on 2D problems: the contact bodies are reduced to plane bodies and the contact surfaces to contact lines.

7.2 Node-To-Node (NTN) Contact Element

We determine, first, the small displacement problem for contact problems by the following assumptions that are valid during the deformation process:

1. Two contacting nodes approximately lie on one normal vector, **n**.
2. Normal vector **n** (and therefore tangent vector τ) changes negligibly.
3. Displacements of two contacting nodes are small in comparison to the size of finite elements used for this discretization.
4. For the discretization of both contacting bodies *only linear finite elements* are used.

If these assumptions are *a priori* fulfilled, then the Node-To-Node contact approach can be used without large loss of tolerance for finite element analysis.

For the Node-To-Node approach, the contact is checked between two nodes (see Figure 13.1 in Chapter 13, Part II), one node is taken on the master body and another node is taken on the slave body. These nodes can be directly taken from the structural finite element mesh of the contacting bodies. Thus, a single NTN contact element consists of a pair of two nodes (master and slave), resulting in a nodal vector of four entries in 2D as

$$\mathbf{x} = \begin{pmatrix} \boldsymbol{\rho} \\ \mathbf{r} \end{pmatrix} = \begin{pmatrix} x_M \\ y_M \\ x_S \\ y_S \end{pmatrix}. \tag{7.5}$$

Very strict assumptions (1) and (2) do not allow us to compute the normal vector, only to specify this as a constant vector during deformation: $\mathbf{n} = \begin{pmatrix} n_x \\ n_y \end{pmatrix}$. The normal vector has to be provided *a priori* for the contact element and has to match approximately with the lines of normals for each body within this area (see Figure 13.1 in Chapter 13, Part II). Based on this discretization the relative displacement vector $\mathbf{r} - \boldsymbol{\rho}$ has the following form

$$\mathbf{r} - \boldsymbol{\rho} = -\begin{bmatrix} 1 & 0 & -1 & 0 \\ 0 & 1 & 0 & -1 \end{bmatrix} \begin{pmatrix} x_M \\ y_M \\ x_S \\ y_S \end{pmatrix} = [A]\mathbf{x} \tag{7.6}$$

where $[A]$ defines the approximation matrix (or position matrix) of the dimension 2×4. The approximation matrix $[A]$, is used for the variation of the relative displacement vector and the relative velocity vector as

$$\delta \mathbf{r} - \delta \boldsymbol{\rho} = [A]\delta \mathbf{x}, \tag{7.7}$$

$$\mathbf{v}_S - \mathbf{v}_M = [A]\mathbf{v}, \tag{7.8}$$

respectively, where $\delta \mathbf{x}$ and $\mathbf{v}$ represent the variations and velocities of the nodal points (finite element nodal vectors).

The penetration computed by equation (4.20) is discretized as follows:

$$\xi^3 = (\mathbf{r} - \boldsymbol{\rho}) \cdot \mathbf{n} = (\mathbf{r} - \boldsymbol{\rho})^T \mathbf{n} = \mathbf{x}^T [A]^T \mathbf{n}. \qquad (7.9)$$

As discussed in Section 7.1.1, the following weak form in equation (7.1) has to be discretized

$$\delta W_N^c = \varepsilon_N H(-\xi^3) \xi^3 \delta \xi^3 = \varepsilon_N H(-\xi^3) \xi^3 \delta \mathbf{x}^T [A]^T \mathbf{n}. \qquad (7.10)$$

This form gives the residual vector after taking away $\delta \mathbf{x}^T$

$$[R] = \varepsilon_N H(-\xi^3) \xi^3 [A]^T \mathbf{n}. \qquad (7.11)$$

The consistent linearization following equations (6.28–6.30) is reduced to the main part only because no rotation and no curvature of the single slave node is possible. The linearization $D_v(\delta W_N^c)$ leads to

$$D_v(\delta W_N^c) = \varepsilon_N H(-\xi^3)(\delta \mathbf{r}_S - \delta \boldsymbol{\rho}) \cdot \mathbf{n} \otimes \mathbf{n} (\mathbf{v}_S - \mathbf{v}_M) \qquad (7.12)$$

$$= \varepsilon_N H(-\xi^3) \delta \mathbf{x}^T [A]^T \mathbf{n} \otimes \mathbf{n} [A] \mathbf{v}. \qquad (7.13)$$

The tangent matrix is derived as the matrix $[K]$ between vectors $\delta \mathbf{x}^T$ and $\mathbf{v}$:

$$[K] = \varepsilon_N H(-\xi^3) [A]^T \mathbf{n} \otimes \mathbf{n} [A]. \qquad (7.14)$$

The tangent matrix $[K]$ is symmetric as would be expected for a conservative system without friction.

Chapter 13 in Part II contains all necessary information for the implementation, verification and further typical examples of the NTN approach based on the Penalty method. Application of other methods, together with NTN discretization, is considered further in Section 14.1, in Chapter 14, Part II for the Lagrange multiplier method and in Chapter 15, in Part II for the Nitsche method.

7.3 Nitsche Node-To-Node (NTN) Contact Element

Within the strong restrictions for the NTN contact element (see Section 7.2) there are some additional particularities for the Nitsche type of contact enforcement.

As before for the penalty formulation, a linearization process such as in equation (6.11) should be applied for equation (6.40) in Section 6.6; thus, it would look like

$$D_v(\delta W_N^c) = D_{\dot{\xi}_I^3} + D_{\delta \dot{\xi}_I^3} + D_{\dot{\sigma}_{N_{II}}} + D_{\delta \dot{\xi}_{II}^3} + D_{\dot{\xi}_{III}^3} + D_{\delta \dot{\sigma}_{N_{III}}}.$$

Hereby, the linearization terms $D_{\delta \dot{\xi}_I^3}$ and $D_{\delta \dot{\xi}_{II}^3}$ vanish as before for the penalty NTN contact element, since no rotation and curvature are possible for the NTN discretization. $D_{\dot{\xi}_I^3}$ gives the same expression as in equation (7.14).

For the further left linearization terms going along with σ_N, we require an insight to the structural part of the contacting bodies. The weak form describing the virtual work for the contacting bodies is necessary:

$$\int_\Omega \sigma_{ij}\, \delta\varepsilon_{ij}\, d\Omega.$$

The discretization of this leads to

$$\delta x_B^T \underbrace{\int_\Omega B^T\, D\, B\, d\Omega}_{ss}\, v_B$$

where ss is the part forming the tangent matrix of a contacting body, B denotes the shape functions in global derivatives and D is the material matrix. Thus, $\dot\sigma_N$ and $\delta\sigma_N$ can be discretized as

$$\dot\sigma_N = \sigma_{ij}\, n_i\, n_j \Rightarrow \dot\sigma_N = n^{2^T} D B\, v_B$$

$$\delta\sigma_N = n^{2^T} D B\, \delta x_B$$

where n^2 enables the normal projection of the stress tensor

$$n^2 = [n_1 n_1,\; n_2 n_2,\; n_3 n_3,\; 2 n_1 n_2,\; 2 n_2 n_3,\; 2 n_1 n_3].$$

Therefore, the linearization terms $D_{\dot\sigma_{N_{II}}}$ and $D_{\dot\xi^3_{III}}$ can be expressed using the analogy between variation and velocity, as

$$D_{\dot\sigma_{N_{II}}} = \delta x_C^T\, H(-\xi^3)\, \underbrace{[a]^T n \otimes n^2 DB}_{scn}\, v_B$$

and

$$D_{\dot\xi^3_{III}} = \delta x_B^T\, H(-\xi^3)\, \underbrace{B^T D^T n^2 \otimes n\, [a]}_{scn^T} v_C$$

where scn and scn^T lead to tangent matrix entries, the indexes B and C express the structural body and contact nodal values, respectively. Furthermore, the last linearization term $D_{\delta\dot\sigma_{N_{III}}}$ vanishes as it can be regarded as a two-times derivative (combination of time derivative and variation) of a linear elastic structural component.

The discretization of equation (6.40) gives the residual entries as

$$\delta W_N^c = \delta x_C^T \left(\underbrace{\varepsilon_N H(-\xi^3)\xi^3 [a]^T n}_{pc} + \underbrace{H(-\xi^3)\sigma_N [a]^T n}_{pcn1} \right)$$

$$+ \delta x_B^T \left(\underbrace{H(-\xi^3)\xi^3 B^T D^T n^2}_{pcn2} \right).$$

As the Nitsche entries are linked to the structural bodies getting in contact, we will use only the master body parts, however, the slave or the average of both bodies would also

Finite Element Discretization

be possible. This has the consequence that v_B and v_C have an overlap $(v_B \cap v_C \neq 0)$. Therefore, a more combined representation seems to be more practicable, including not only the contact entries but also the entries of the master body as

$$\begin{bmatrix} ss & scn^T \\ scn & sc \end{bmatrix} \begin{bmatrix} v_B \\ v_C \end{bmatrix} = \begin{bmatrix} ps + pcn2 \\ pc + pcn1 \end{bmatrix}$$

where $ps = DBv_B$ is the residual vector for the linear elastic structural body necessary within a nonlinear solution scheme and sc is the part of the tangent matrix going along with equation (7.14).

Chapter 15 contains all necessary information for the implementation, verification and further examples of the Nitsche NTN approach based on structural truss elements. Thereby, the contact stress contributions n^2DB get simplified for trusses to $EA\,B$ where EA represents the stiffness of the truss element.

7.4 Node-To-Segment (NTS) Contact Element

The Node-To-Segment contact element appears to be the most popular contact approach in many software packages for finite element analysis. In this case only one of the very strict assumptions for NTN element (see Section 7.2), is left:

- For discretization of the slave part, only finite elements with linear approximations are used.

This requirement will be explained later and is not necessary for the Segment-To-Segment (STS) approach. Both large deformations for bodies and large relative sliding between contacting bodies are allowed. In general, the master surface can be discretized with arbitrary shape functions. It is possible to use NTS contact approach if the slave part is arbitrary discretized, but is not deformable (rigid). In this case the contact slave boundary is described by a set of fixed slave nodes. The Node-To-Segment (NTS) approach performs as a robust element for large displacement problems, even when large sliding between the contacting bodies is ensured. That is why the NTS approach remain the most applicable in explicit finite element packages for linear finite elements.

A single NTS contact element is given by a set of n nodes for discretization of the master surface (e.g., $n = 2$ for linear approximation) and a separated slave node $n + 1$, on from the slave surface. The nodal vector for a NTS element is given as

$$\mathbf{x} = \begin{pmatrix} \boldsymbol{\rho} \\ \mathbf{r} \end{pmatrix} = \begin{pmatrix} x_M^{(1)} \\ y_M^{(1)} \\ \vdots \\ x_M^{(n)} \\ y_M^{(n)} \\ x_S \\ y_S \end{pmatrix}. \tag{7.15}$$

The relative displacement vector $\mathbf{r} - \boldsymbol{\rho}(\xi^1)$ using the shape functions $N^{(k)}(\xi^1)$, $k = 1, 2,...,n$ for the approximation of the master surface can be expressed as follows

$$\mathbf{r} - \boldsymbol{\rho}(\xi^1) = -\begin{bmatrix} N^{(1)} & 0 & \cdots & N^{(n)} & 0 & -1 & 0 \\ 0 & N^{(1)} & \cdots & 0 & N^{(n)} & 0 & -1 \end{bmatrix} \begin{pmatrix} x_M^{(1)} \\ y_M^{(1)} \\ \vdots \\ x_M^{(n)} \\ y_M^{(n)} \\ x_S \\ y_S \end{pmatrix}$$

$$= [A(\xi^1)]\, \mathbf{x} \qquad (7.16)$$

where the approximation matrix $[A(\xi^1)]$ has the dimension of $2 \times 2(n+1)$. For the linear NTS contact element with Lagrange shape functions as shown in Figure 7.1(b) and in Figure 7.2, the approximation matrix is written as

$$[A(\xi^1)] = -\begin{bmatrix} \dfrac{1-\xi^1}{2} & 0 & \dfrac{1+\xi^1}{2} & 0 & -1 & 0 \\ 0 & \dfrac{1-\xi^1}{2} & 0 & \dfrac{1+\xi^1}{2} & 0 & -1 \end{bmatrix} \qquad (7.17)$$

The derivative of the approximation matrix is given as

$$\left[\frac{\partial A}{\partial \xi^1}\right] = [A_{\xi^1}] = \begin{bmatrix} \tfrac{1}{2} & 0 & -\tfrac{1}{2} & 0 & 0 & 0 \\ 0 & \tfrac{1}{2} & 0 & -\tfrac{1}{2} & 0 & 0 \end{bmatrix} \qquad (7.18)$$

This matrix is used to discretize the tangent vector, taking derivative of equation (7.16)

$$\boldsymbol{\rho}_{\xi^1} = -[A_{\xi^1}]\, \mathbf{x} = \begin{bmatrix} -\tfrac{1}{2} & 0 & \tfrac{1}{2} & 0 & 0 & 0 \\ 0 & -\tfrac{1}{2} & 0 & \tfrac{1}{2} & 0 & 0 \end{bmatrix} \mathbf{x}$$

$$= \frac{\mathbf{x}_M^{(2)} - \mathbf{x}_M^{(1)}}{2}. \qquad (7.19)$$

The normal vector $\mathbf{n}$ in 2D case, however, can not be uniquely defined based only on 2D geometry and requires additional definition depending on the modeling guaranteeing that the normal vector $\mathbf{n}$ is pointing outside of the continuum of the master body, see Figure 7.1(b). The linear NTS contact element is defined by two master nodes 1 and 2 and the third slave node. The inner part of the master body lays at the right part from the line connecting nodes 1 and 2, see Figure 7.1(b) and Figure 7.2. The unit tangent vector $\boldsymbol{\tau}$ is calculated from equation (7.19) as normalized vector $\boldsymbol{\rho}_{\xi^1}$

$$\boldsymbol{\tau} = \frac{\boldsymbol{\rho}_{\xi^1}}{|\boldsymbol{\rho}_{\xi^1}|} = \frac{\mathbf{x}^{(2)} - \mathbf{x}^{(1)}}{|\mathbf{x}^{(2)} - \mathbf{x}^{(1)}|} = \frac{1}{|\mathbf{x}^{(2)} - \mathbf{x}^{(1)}|} \begin{bmatrix} x^{(2)} - x^{(1)} \\ y^{(2)} - y^{(1)} \end{bmatrix}. \qquad (7.20)$$

Finite Element Discretization

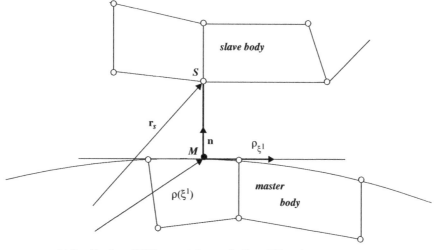

(a) Specification of NTS contact element for linear FE mesh

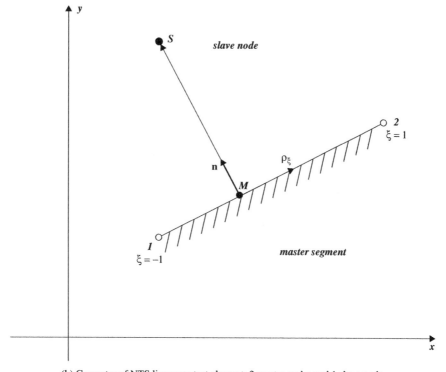

(b) Geometry of NTS linear contact element: 2 master nodes and 1 slave node

Figure 7.1 Geometry and kinematics of the Node-To-Segment contact element

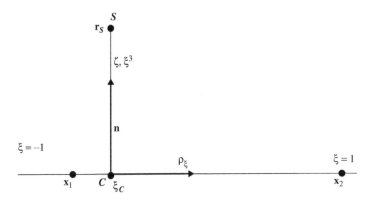

Figure 7.2 Node-To-Segment (NTS) contact element. The lower part is the inner part of the body. Coordinate system ρ_ξ, $\mathbf{n}$ and coordinates ζ, ξ

The normal vector $\mathbf{n}$ is defined via the cross product with the third Cartesian basis vector $\mathbf{k}$ along z-axis

$$\mathbf{n} = \mathbf{k} \times \boldsymbol{\tau} = \begin{vmatrix} i & j & k \\ 0 & 0 & 1 \\ x^{(2)} - x^{(1)} & y^{(2)} - y^{(1)} & 0 \end{vmatrix} \frac{1}{|\mathbf{x}^{(2)} - \mathbf{x}^{(1)}|}$$

$$= \frac{1}{|\mathbf{x}^{(2)} - \mathbf{x}^{(1)}|} \begin{bmatrix} y^{(1)} - y^{(2)} \\ x^{(2)} - x^{(1)} \end{bmatrix}. \tag{7.21}$$

7.4.1 Closest Point Projection Procedure for the Linear NTS Contact Element

In order to setup the local coordinate system on the master segment to describe all contact parameters (Section 4.1), it is necessary first to solve the CPP procedure (Section 4.2.2). The solution will give us a projection point $\mathbf{C}$, namely its coordinate $\xi^1 = \xi_C$ (see Figure 7.2). Evaluating equation (4.14) for the increment $\Delta\xi^1_{(n)}$ together with the derivatives for the linear NTS element in equation (4.15) and taking into account that the second derivative of the master position vector $\rho_{\xi^1\xi^1}$ becomes zero, we obtain the following expression

$$\Delta\xi^1_{(n)} = \frac{(\mathbf{r} - \boldsymbol{\rho}) \cdot \boldsymbol{\rho}_{\xi^1}}{\boldsymbol{\rho}_{\xi^1} \cdot \boldsymbol{\rho}_{\xi^1} - (\mathbf{r} - \boldsymbol{\rho}) \cdot \boldsymbol{\rho}_{\xi^1\xi^1}}$$

$$= \frac{4}{\left|\mathbf{x}_M^{(2)} - \mathbf{x}_M^{(1)}\right|^2} \left[\mathbf{r} - \frac{1-\xi^1}{2}\mathbf{x}_M^{(1)} - \frac{1+\xi^1}{2}\mathbf{x}_M^{(2)} \right] \cdot \frac{\mathbf{x}_M^{(2)} - \mathbf{x}_M^{(1)}}{2}$$

$$= \frac{4}{\left|\mathbf{x}_M^{(2)} - \mathbf{x}_M^{(1)}\right|^2} \left[\mathbf{r} \cdot \frac{\mathbf{x}_M^{(2)} - \mathbf{x}_M^{(1)}}{2} - \frac{1}{4}(\mathbf{x}_M^{(2)} + \mathbf{x}_M^{(1)}) \cdot (\mathbf{x}_M^{(2)} - \mathbf{x}_M^{(1)}) - \right.$$

Finite Element Discretization

$$-\frac{1}{4}\xi^1 \left|\mathbf{x}_M^{(2)} - \mathbf{x}_M^{(1)}\right|^2 \Bigg]$$

$$= -\xi^1 + \frac{2\mathbf{r} \cdot (\mathbf{x}_M^{(2)} - \mathbf{x}_M^{(1)}) - \left|\mathbf{x}_M^{(2)}\right|^2 + \left|\mathbf{x}_M^{(1)}\right|^2}{\left|\mathbf{x}_M^{(2)} - \mathbf{x}_M^{(1)}\right|^2}. \tag{7.22}$$

This leads, for the Newton iterative solution scheme in equation (4.14), to a single iteration

$$\xi_C = \xi^1_{(n+1)} = \xi^1_{(n)} + \Delta\xi^1_{(n)}$$

$$= \frac{2\mathbf{r}_S \cdot (\mathbf{x}_M^{(2)} - \mathbf{x}_M^{(1)}) - |\mathbf{x}_M^{(2)}|^2 + |\mathbf{x}_M^{(1)}|^2}{|\mathbf{x}_M^{(2)} - \mathbf{x}_M^{(1)}|^2}. \tag{7.23}$$

This result represents a closed form solution of the CPP procedure for a linearly approximated master segment in 2D, since the solution does not depend on the value ξ^1 from the previous iteration step. Existence of the simple closed form solution reflects the simple triangular geometry of the linear NTS contact element.

Thus, using a local coordinate system for a linear element only as

$$\mathbf{r}_S = \mathbf{x}^{(1)} + \boldsymbol{\tau}\xi^{(1)} + \mathbf{n}\xi^{(3)} \tag{7.24}$$

with the constant unit tangent vector $\boldsymbol{\tau}$ in equation (7.20) and the constant unit normal vector $\mathbf{n}$ in equation (7.21), the convective coordinates can be immediately obtained via the scalar product

$$\xi^1 = (\mathbf{r}_S - \mathbf{x}^{(1)}) \cdot \boldsymbol{\tau} \tag{7.25}$$

$$\xi^3 = (\mathbf{r}_S - \mathbf{x}^{(1)}) \cdot \mathbf{n} \tag{7.26}$$

However, is recommended in computation to use the general expression for penetration via equation (7.9) as

$$\xi^3 = (\mathbf{r}_S - \boldsymbol{\rho}) \cdot \mathbf{n}. \tag{7.27}$$

7.4.2 Peculiarities in Computation of the Contact Integral

Computation of the contact integral within the NTS approach is performed at the projection point ξ_C after the CPP procedure and looks similar to the NTN approach because the slave boundary is represented by a single node (see equation (7.2) and discussion in Section 7.1.2). This procedure can be interpreted as a Lobatto type of integration (see Appendix A), with two integration points as

$$\int_{-1}^{1} f(\xi^1) d\xi^1 \approx f(1) + f(-1) \tag{7.28}$$

where $\xi_i^1 = \{-1, 1\}$ are the integration points and $w_i = \{-1, 1\}$ the corresponding weights.

Remark 7.4.1 (Interpretation of NTS as Lobatto type integration) *The NTS approach with a linearly approximated slave segment can be interpreted as numerical integration with two Lobatto integration points coinciding with the slave nodes and leading to two contact elements with the same master segment. Since the Lobatto formula with two integration points allows us to integrate exactly only the linear function (see Appendix A) the application of NTS approach to finite elements with approximations an order higher than linear will lead to large errors in computation of contact stresses. In other words, in order to correctly transfer uniform contact pressures from the slave to master segment even the NTS approach is limited to only a linear approximation of the slave segment.*

Remark 7.4.2 (Selection of master and slave parts for NTS approach) *It is also necessary to choose the size of contacting elements for master and slave correctly in order to transfer the stresses correctly. Namely, the slave segment must be smaller in size than the master segment, because the opposite choice – master segment smaller than slave – leads to loss of contact. A special patch test (see Section 18.2.3 in Part II) is configured to check the correct transfer of contact stresses and will be studied numerically.*

The two-path contact algorithm, including a double check from both master and slave parts (they change their roles), is implemented in many finite element codes in order to overcome the patch test problem and be free from the special choice for the master and slave parts.

7.4.3 Residual and Tangent Matrix

The weak form in equation (7.2) is discretized using the approximation matrix $[A]$ in equation (7.17)

$$\delta W_N^c = \varepsilon_N H(-\xi^3)\xi^3 \delta\xi^3$$
$$= \varepsilon_N H(-\xi^3)\xi^3 (\delta\mathbf{r} - \delta\boldsymbol{\rho}(\xi^1)) \cdot \mathbf{n}$$
$$= \varepsilon_N H(-\xi^3)\xi^3 \delta\mathbf{x}^T [A]^T \mathbf{n}, \qquad (7.29)$$

which gives the following residual after taking away the vector $\delta\mathbf{x}^T$

$$[R] = \varepsilon_N H(-\xi^3)\xi^3 [A]^T \mathbf{n}. \qquad (7.30)$$

The tangent matrix for the current linear NTS contact element (in addition $[A_\xi]$ in equation (7.18) is employed) is derived as after the discretization of equations (6.28–6.29) as

$$D_v(\delta W_N^c) = \varepsilon_N H(-\xi^3)\{(\delta\mathbf{r} - \delta\boldsymbol{\rho}) \cdot (\mathbf{n}\otimes\mathbf{n})(\mathbf{v}_S - \mathbf{v}_M) \qquad (7.31)$$
$$-\xi^3 [\delta\boldsymbol{\rho}_{\xi^1}\cdot(\mathbf{n}\otimes\boldsymbol{\rho}_{\xi^1})(\mathbf{v}_S - \mathbf{v}_M)m^{11} + (\delta\mathbf{r} - \delta\boldsymbol{\rho})\cdot(\boldsymbol{\rho}_{\xi^1}\otimes\mathbf{n})\mathbf{v}_{M_{\xi^1}} m^{11}]\}$$

$$= \varepsilon_N H(-\xi^3)\{\delta\mathbf{x}^T[A]^T\mathbf{n} \otimes \mathbf{n}[A]\mathbf{v}$$
$$-\xi^3 m^{11}(\delta\mathbf{x}^T[A_{\xi^1}]^T\mathbf{n} \otimes \boldsymbol{\rho}_{\xi^1}[A]\mathbf{v} + \delta\mathbf{x}^T[A]^T\boldsymbol{\rho}_{\xi^1} \otimes \mathbf{n}[A_{\xi^1}]\mathbf{v})\}.$$

Here, the metric component

$$m^{11} = \frac{1}{\boldsymbol{\rho}_{\xi^1} \cdot \boldsymbol{\rho}_{\xi^1}}, \qquad (7.32)$$

(see Remark 4.3.1) appears as a result of transformation from the unit tangent τ to the tangent vector $\boldsymbol{\rho}_{\xi^1}$. The tangent matrix is computed as a part of equation (7.31) between vectors $\delta\mathbf{x}^T$ and $\mathbf{v}$:

$$[K] = \varepsilon_N H(-\xi^3)\{[A]^T\mathbf{n} \otimes \mathbf{n}[A] \qquad (7.33)$$
$$-\xi^3 m^{11}([A_{\xi^1}]^T\mathbf{n} \otimes \boldsymbol{\rho}_{\xi^1}[A] + [A]^T\boldsymbol{\rho}_{\xi^1} \otimes \mathbf{n}[A_{\xi^1}]\}. \qquad (7.34)$$

The tangent matrix $[K]$ is symmetric as expected for a conservative system without friction – the second line (7.34) contains the rotational part in the form a sum of the arbitrarily matrix and it transposition: $[C] + [C]^T$, which leads to the symmetric matrix.

Remark 7.4.3 *Both the residual and the tangent matrix have to be computed at the projection point ($\xi^1 = \xi_C$) derived by the CPP procedure.*

Remark 7.4.4 *Both contact bodies can be deformable, but the slave body has to be necessarily linearly approximated for being rigid and fixed.*

Remark 7.4.5 (Necessity of the main part) *The main part in equation (7.33) plays the major role during the iterative solution. The correct contact kinematics (non-penetration) will be observed already with the normal part only without implementation of the rotational part.*

Chapter 16 in Part II contains all necessary information for the implementation, verification and further examples for the NTS non-frictional contact element.

Remark 7.4.6 (Meaning of the rotational part) *For both cases, either N is an independent variable, such as for the Lagrange multipliers method, or N is given external normal force (following forces case); only the rotational part is remains for the linear segment (plus curvature part for the curvilinear segment). A special example of the following normal force, based on NTS implementation as an inverted contact algorithm, is considered in Example 16.3, Chapter 16 in Part II.*

In this book, we only show peculiarities within the implementation of contact elements for the 3D case for the NTS approach. Chapter 20 in Part II contains all necessary information for the implementation, verification and further examples for the 3D NTS non-frictional contact element. At that point, for further 3D implementation the advanced reader is referred to the research monograph by Konyukhov and Schweizerhof (2012) as well as to other famous books on computational contact mechanics see Wriggers, 2002.

The NTS approach is widely used for dynamic problems, especially for explicit simulations (LS-DYNA, Abaqus FEA programs). Application to both implicit and explicit dynamic simulation is straightforward, however, we have to mention that there is a separate set of problems for dynamic simulation, such as conservation of momentum, conservation of energy and selection of the time step. All these require special study of dynamic contact problems. Nevertheless, we show dynamic implementation in its simplest NTS implementation in Chapter 24 Transient contact problems together with the verification of the specific center of percussion problem, solved analytically in Section 8.3.

7.5 Segment-To-Analytical-Surface (STAS) Approach

Contact problems between a deformable and a rigid body, as described for the Signorini problem in Chapter 2, can be modeled using the Segment-To-Analytical-Surface approach. The approach is based on the simplicity of computation of normal vectors to the analytically given surface. Using the master-slave terminology, the deformable body is denoted the slave body and the rigid body the master body. This allows easy computation of penetration. The rigid body is, however, not discretized but described analytically by a surface function in 3D or a curve in 2D. Although the rigid surface is denoted the master surface for the computation of penetration, the relative coordinate system for the contact kinematics description is set on the slave body as well, see Figures 7.3 and 7.4. The contact is checked between the slave deformable segment, described by a set of integration points and an analytical surface. In general, the case of the CPP procedure is involved to determine the position α on the rigid surface/curve where an integration point from the slave segment (FE mesh) penetrates. In addition, one can distinguish two possibilities to compute penetration: either in the coordinate system of a master segment or in the coordinate system of a slave segment, see more in the book by Konyukhov and Schweizerhof (2012).

7.5.1 General Structure of CPP Procedure for STAS Contact Element

For the analytically given curve

$$\mathbf{r}_s(\alpha) = \begin{Bmatrix} x(\alpha) \\ y(\alpha) \end{Bmatrix} \tag{7.35}$$

and the given integration point ξ^1 on the slave segment, the CPP procedure is defined for each given slave point ξ^1 as the minimization problem with respect to variable α

$$\|\boldsymbol{\rho}(\xi^1) - \mathbf{r}_s(\alpha)\| \to \min. \tag{7.36}$$

Finite Element Discretization

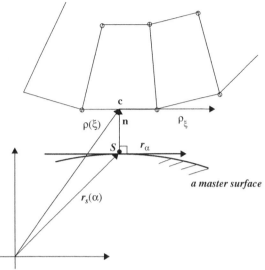

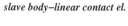

Figure 7.3 Geometry and kinematics of the Segment-To-Analytical-Surface (STAS)

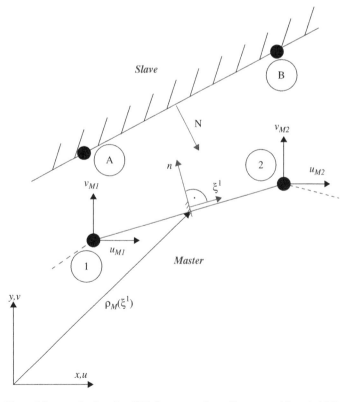

Figure 7.4 Closed form solution for STAS penetration: Contact with a rigid line. Coordinate system for computation of the penetration

For the solution we need the optimality condition

$$F'(\alpha) = (\rho(\xi^1) - \mathbf{r}_s(\alpha)) \cdot \mathbf{r}_s(\alpha)_\alpha = 0. \tag{7.37}$$

Equation (7.37) can be solved by the Newton method in full analogy to the result discussed in Section 4.2.2, but now with regard to the variable α. The result of the CPP procedure for integration points distributed along the slave contacting curve (surface) is used to compute the penetration into the rigid master curve (surface). The computation of penetration for some particular cases of simple rigid body geometries is also possible in the closed form. The closed form solution is possible for the following rigid body surfaces in 3D: plane, cylinder, sphere and torus; and for lines and circles in 2D.

7.5.2 Closed form Solutions for Penetration in 2D

7.5.2.1 Master Surface as an Analytical Rigid Straight Line

We consider a line representing the boundary of a rigid body going through the points A and B, see Figure 7.4:

$$\mathbf{r}_S = \mathbf{x}^{(A)} + \eta \boldsymbol{\tau}^{(AB)} \tag{7.38}$$

where η is parametrization of the line and $\boldsymbol{\tau}^{(AB)}$ is the unit tangent of the line or just a unit vector pointing from $\mathbf{A}$ to $\mathbf{B}$. On the other hand, the line can also be described by the local coordinate system for the contact element with nodes 1 and 2 (see Figure 7.4) as

$$\mathbf{r}_S = \rho(\xi^1) + \xi^3 \mathbf{n} \tag{7.39}$$

where $\mathbf{n}$ denotes the normal on the contact element, which is constructed as

$$\mathbf{n} = \frac{1}{|\mathbf{x}^{(2)} - \mathbf{x}^{(1)}|} \begin{pmatrix} y_1 - y_2 \\ x_2 - x_1 \end{pmatrix}. \tag{7.40}$$

Both equations (7.38) and (7.39) represent the same line $\mathbf{r}_S$, therefore, the subtraction gives us

$$\rho(\xi^1) + \xi^3 \mathbf{n} = \mathbf{x}^{(A)} + \eta \boldsymbol{\tau}^{(AB)}. \tag{7.41}$$

Taking now the scalar product with the normal vector $\mathbf{N}$ to the rigid line (orthogonal to $\boldsymbol{\tau}^{(AB)}$) in order to exclude operations with unknown η, we obtain the following expression

$$\rho(\xi^1) \cdot \mathbf{N} + \xi^3 \mathbf{n} \cdot \mathbf{N} = \mathbf{x}^{(A)} \cdot \mathbf{N}. \tag{7.42}$$

The last equation is resolved via the penetration:

$$\xi^3 = \frac{(\mathbf{x}^{(A)} - \rho(\xi^1)) \cdot \mathbf{N}}{\mathbf{n} \cdot \mathbf{N}}. \tag{7.43}$$

Finite Element Discretization

The normal vector $\mathbf{N}$ on the rigid part can be written in due course as

$$\mathbf{N} = \frac{1}{|\mathbf{x}^{(A)} - \mathbf{x}^{(B)}|} \begin{pmatrix} y_B - y_A \\ x_A - x_B \end{pmatrix}. \tag{7.44}$$

Equation (7.43) gives the penetration in closed form for the STAS discretization of a rigid body with a straight line as a boundary. The formula is restricted to bodies with straight contact boundary lines not being orthogonal to each other ($\mathbf{n} \cdot \mathbf{N} \neq 0$).

7.5.2.2 Master Surface as Analytical Rigid Circle

The computation of the penetration is even more simple for the analytical rigid circle as shown in Figure (7.5). It is necessary to distinguish whether the rigid circular line describes the inner or outer part of the circle.

If the inner part of the circle is the rigid body then the penetration is computed as

$$\xi^3 = |\mathbf{r}_0 - \boldsymbol{\rho}(\xi^1)| - R; \tag{7.45}$$

If the outer part of the circle is the rigid body then the penetration is computed as

$$\xi^3 = R - |\mathbf{r}_0 - \boldsymbol{\rho}(\xi^1)|, \tag{7.46}$$

respectively. $\mathbf{r}_0$ denotes the center and R the radius of the circle.

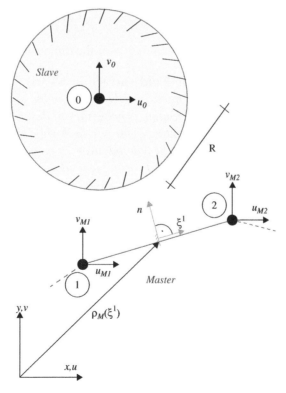

Figure 7.5 STAS: Contact with a rigid circle. Coordinate system for computation of the penetration

7.5.3 Discretization for STAS Contact Approach

A single STAS contact element consists of a pair of n nodes for the discretization of the slave segment (e.g., $n = 2$ for linear approximation). The resulting nodal vector for the STAS contact element is expressed by

$$\mathbf{x} = (\rho) = \begin{pmatrix} x_M^{(1)} \\ y_M^{(1)} \\ \vdots \\ x_M^{(n)} \\ y_M^{(n)} \end{pmatrix}. \tag{7.47}$$

Based on this discretization of the slave segment, and using the shape functions $N^{(k)}(\xi^1)$, $k = 1, 2,...,n$, the relative displacement vector $\mathbf{r} - \rho(\xi^1)$ has the following form

$$\mathbf{r} - \rho(\xi^1) = -\begin{bmatrix} N^{(1)} & 0 & \cdots & N^{(n)} & 0 \\ 0 & N^{(1)} & \cdots & 0 & N^{(n)} \end{bmatrix} \begin{pmatrix} x_M^{(1)} \\ y_M^{(1)} \\ \vdots \\ x_M^{(n)} \\ y_M^{(n)} \end{pmatrix} = [A(\xi^1)]\mathbf{x} \tag{7.48}$$

where $[A(\xi^1)]$ defines the approximation matrix of the dimension $2 \times 2n$.

Remark 7.5.1 (Description of the rigid surface in STAS) *The rigid boundary described by analytical functions is not explicitly written in the approximation matrix $[A(\xi^1)]$, however, an additional subroutine is necessary to describe the rigid surface for all contact segments. This subroutine – one for each rigid surface and common for all contact segments – includes the computation of penetration either in the general form as described in Section 7.5.1, or in particular closed form as described in Sections. 7.5.2.1–7.5.2.2.*

7.5.4 Residual and Tangent Matrix

Finite element discretization for the weak form follows the numerical integration discussed in Section 7.1.3 with regard to the approximation matrix given in equation (7.48). Equation (7.3) is transformed as follows:

$$\delta W_N^c = \int_S \varepsilon_N H(-\xi^3) \xi^3 \delta \xi^3 \, ds$$

$$= \int_S \varepsilon_N H(-\xi^3) \xi^3 (-\delta\rho) \cdot \mathbf{n} \, ds$$

$$= \int_{\xi^1} \varepsilon_N H(-\xi^3) \xi^3 \delta \mathbf{x}^T [A]^T \mathbf{n} \sqrt{\rho_{\xi^1} \cdot \rho_{\xi^1}} \, d\xi^1$$

›
Finite Element Discretization

$$= \sum_{q=1}^{N} \varepsilon_N H(-\xi^3)\xi^3 \delta \mathbf{x}^T [A]^T \mathbf{n} \sqrt{\boldsymbol{\rho}_{\xi^1} \cdot \boldsymbol{\rho}_{\xi^1}} \, w_q, \qquad (7.49)$$

where various numerical integration rules can be applied (see Appendix A). Equation (7.49) leads to the following residual (after taking away $\delta \mathbf{x}^T$)

$$[R] = \sum_{q=1}^{N} \varepsilon_N H(-\xi^3)\xi^3 [A]^T \mathbf{n} \sqrt{\boldsymbol{\rho}_{\xi^1} \cdot \boldsymbol{\rho}_{\xi^1}} \, w_q. \qquad (7.50)$$

The tangent matrix for the linear approximation is derived with regard to the discretization of the consistent linearization in equations (6.28–6.29):

$$D_v(\delta W_N^c) = \int_S \varepsilon_N H(-\xi^3)\{-\delta\boldsymbol{\rho} \cdot (\mathbf{n} \otimes \mathbf{n})(\mathbf{v}_S - \mathbf{v}_M)$$

$$-\xi^3 [\delta\boldsymbol{\rho}_{\xi^1} \cdot (\mathbf{n} \otimes \boldsymbol{\rho}_{\xi^1})(\mathbf{v}_S - \mathbf{v}_M) m^{11} + (-\delta\boldsymbol{\rho}) \cdot (\boldsymbol{\rho}_{\xi^1} \otimes \mathbf{n}) \mathbf{v}_{M_{\xi^1}} m^{11}]\} \, ds$$

$$= \sum_{q=1}^{N} \varepsilon_N H(-\xi^3)\{\delta \mathbf{x}^T [A]^T \mathbf{n} \otimes \mathbf{n}[A]\mathbf{v} \qquad (7.51)$$

$$-\xi^3 m^{11}(\delta \mathbf{x}^T [A_{\xi^1}]^T \mathbf{n} \otimes \boldsymbol{\rho}_{\xi^1}[A]\mathbf{v} + \delta \mathbf{x}^T [A]^T \boldsymbol{\rho}_{\xi^1} \otimes \mathbf{n}[A_{\xi^1}]\mathbf{v})\}$$

$$\times \sqrt{\boldsymbol{\rho}_{\xi^1} \cdot \boldsymbol{\rho}_{\xi^1}} \, w_q.$$

The tangent matrix is derived as a matrix $[K]$ between $\delta \mathbf{x}^T$ and v in equation (7.51):

$$[K] = \sum_{q=1}^{N} \varepsilon_N H(-\xi^3)\{[A]^T \mathbf{n} \otimes \mathbf{n}[A] \qquad (7.52)$$

$$-\xi^3 m^{11}([A_{\xi^1}]^T \mathbf{n} \otimes \boldsymbol{\rho}_{\xi^1}[A] + [A]^T \boldsymbol{\rho}_{\xi^1} \otimes \mathbf{n}[A_{\xi^1}])\} \sqrt{\boldsymbol{\rho}_{\xi^1} \cdot \boldsymbol{\rho}_{\xi^1}} \, w_q. (7.53)$$

Chapter 17 in Part II contains all necessary information for the implementation, verification and further examples for the STAS non-frictional contact element.

Remark 7.5.2 (Inverted contact algorithm and following forces) *Implementation of the rotational part (plus curvature part in case of high-order FE approximations) with the treatment of N as an external forces (inverted contact algorithm) gives us a general case of implementation of the following normal force case. A case of a single following force is considered in Example 17.3.1 (Chapter 17 in Part II) and a case of distributed forces is considered in Example 17.3.2, and for inflation in Example 17.3.3, (Chapter 17 in Part II).*

7.6 Segment-To-Segment (STS) Mortar Approach

The Segment-To-Segment, or sometimes Surface-To-Surface (STS) Mortar type discretization, in comparison with the NTS approach, is more universal because:

- it allows us to use arbitrary size discretization independently for both master and slave parts, therefore satisfying the patch test (see Remark 7.4.2);
- it allows us to use any order of approximations (inclusive high-order and iso-geometric finite element techniques) for both master and slave parts.

A single STS contact element consists of n nodes for the discretization of the master side and of m nodes for the slave surface (e.g., $n = 2, m = 2$ for linear approximation of master and slave surfaces) (see Figure 7.6).

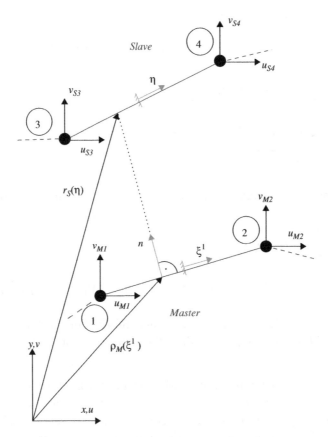

Figure 7.6 Geometry and kinematics of the Segment-To-Segment contact element

Finite Element Discretization

The nodal vector for the STS contact element in 2D is written as

$$\mathbf{x} = \begin{pmatrix} \boldsymbol{\rho} \\ \mathbf{r} \end{pmatrix} = \begin{pmatrix} x_M^{(1)} \\ y_M^{(1)} \\ \vdots \\ x_M^{(n)} \\ y_M^{(n)} \\ x_S^{(1)} \\ y_S^{(1)} \\ \vdots \\ x_S^{(m)} \\ y_S^{(m)} \end{pmatrix}. \tag{7.54}$$

Approximating both the master and slave contact boundaries with the shape functions $N^{(k)}(\xi^1), k = 1, 2,...,n$ and $M^{(l)}(\eta), l = 1, 2,...m$, the relative displacement vector can be expressed as

$$\mathbf{r}(\eta) - \boldsymbol{\rho}(\xi^1) = \tag{7.55}$$

$$= \begin{bmatrix} -N^{(1)} & 0 & \cdots & -N^{(n)} & 0 & M^{(1)} & 0 & \cdots & M^{(m)} & 0 \\ 0 & -N^{(1)} & \cdots & 0 & -N^{(n)} & 0 & M^{(1)} & \cdots & 0 & M^{(m)} \end{bmatrix} \begin{pmatrix} x_M^{(1)} \\ y_M^{(1)} \\ \vdots \\ x_M^{(n)} \\ y_M^{(n)} \\ x_S^{(1)} \\ y_S^{(1)} \\ \vdots \\ x_S^{(m)} \\ y_S^{(m)} \end{pmatrix}$$

$$= [A(\xi^1, \eta)] \, \mathbf{x}$$

with the approximation matrix $[A(\xi^1, \eta)]$ possessing the dimension of $2 \times 2(n+m)$.

The approximation matrix for the STS contact element with linear Lagrange shape functions for both master and slave sides ($n = m = 2$) is written as

$$[A(\xi^1, \eta)] = \begin{bmatrix} -\frac{1-\xi^1}{2} & 0 & -\frac{1+\xi^1}{2} & 0 & \frac{1-\eta}{2} & 0 & \frac{1+\eta}{2} & 0 \\ 0 & -\frac{1-\xi^1}{2} & 0 & -\frac{1+\xi^1}{2} & 0 & \frac{1-\eta}{2} & 0 & \frac{1+\eta}{2} \end{bmatrix}. \tag{7.56}$$

The tangent vectors ρ_{ξ^1} and $\mathbf{r}_\eta$ on the master and slave surface are computed, respectively, as

$$\rho_{\xi^1} = -[A_{\xi^1}]\,\mathbf{x} = \frac{\mathbf{x}_M^{(2)} - \mathbf{x}_M^{(1)}}{2} \qquad (7.57)$$

and

$$\mathbf{r}_\eta = [A_\eta]\,\mathbf{x} = \frac{\mathbf{x}_S^{(2)} - \mathbf{x}_S^{(1)}}{2}, \qquad (7.58)$$

where the corresponding derivatives of the matrix $[A(\xi^1, \eta)]$ are involved:

$$[A_{\xi^1}] = \frac{\partial [A]}{\partial \xi^1} = \frac{1}{2}\begin{bmatrix} 1 & 0 & -1 & 0 & 0 & 0 & 0 & 0 \\ 0 & 1 & 0 & -1 & 0 & 0 & 0 & 0 \end{bmatrix}. \qquad (7.59)$$

$$[A_\eta] = \frac{\partial [A]}{\partial \eta} = \frac{1}{2}\begin{bmatrix} 0 & 0 & 0 & 0 & -1 & 0 & 1 & 0 \\ 0 & 0 & 0 & 0 & 0 & -1 & 0 & 1 \end{bmatrix}. \qquad (7.60)$$

The normal vector to the master segment $\mathbf{n}(\xi^1)$ is defined in the direction outside the continuum of the master body (similar to the NTS discretization, see equation (7.21)).

7.6.1 Peculiarities of the CPP Procedure for the STS Contact Approach

The CPP procedure is required to determine convective coordinates ξ_i^1 on the master segment as a result of the projection of each integration point η_i distributed at the slave segment (see Figure 7.6). This allows us to derive a set of active points (see discussion in Section 7.1.3) in order to compute further both the residual and the tangent matrices. Though the CPP procedure requires the general iterative solution, in the case of linear approximation for both slave and master segments ($[A]$ is given by equation (7.56)), the closed form solution in equation (7.23) is applicable to compute ξ_i^1 as discussed for NTS discretization.

7.6.2 Computation of the Residual and Tangent Matrix

Numerical integration for the STS approach is performed by including the active set of slave integration points as discussed in Section 7.1.4. Finite element discretization of equation (7.4) leads to the following

$$\delta W_N^c = \sum_{q=1}^{N} \varepsilon_N H(-\xi^3)\xi^3 (\delta \mathbf{r}(\eta) - \delta \rho(\xi^1)) \cdot \mathbf{n}\,\sqrt{\mathbf{r}_\eta \cdot \mathbf{r}_\eta}\, w_q$$

$$= \sum_{q=1}^{N} \varepsilon_N H(-\xi^3)\xi^3 \delta\mathbf{x}^T [A(\xi^1, \eta)]^T \mathbf{n}\,\sqrt{\mathbf{r}_\eta \cdot \mathbf{r}_\eta}\, w_q \qquad (7.61)$$

Finite Element Discretization

resulting in the residual

$$\{R\} = \sum_{q=1}^{N} \varepsilon_N H(-\xi^3)\xi^3 [A]^T \mathbf{n} \sqrt{\mathbf{r}_\eta \cdot \mathbf{r}_\eta}\, w_q, \qquad (7.62)$$

where the expression

$$\sqrt{\mathbf{r}_\eta \cdot \mathbf{r}_\eta} = \frac{\|\mathbf{x}_S^{(2)} - \mathbf{x}_S^{(1)}\|}{2} \qquad (7.63)$$

appears as the computation of the arc-length $ds(\eta)$ (Jacobian of transformation $s \to \eta$), see equation (3.2) in Chapter 3 on differential geometry.

Remark 7.6.1 *Depending on the goals, it is possible to use various integration formulas (Gauss or Lobatto) to cover the slave segment. As is known, the number of integration points can not be defined a priori, because the contact integrals are no longer smooth, or even continuous functions on the given segment. Application of the complex formula with subdivisions can be considered an optimal solution in comparison to a simple increase in the number of integration points.*

The tangent matrix for the current linear STS contact element is computed as the matrix $[K]$ between vectors $\delta\mathbf{x}^T$ and $\mathbf{v}$:

$$D_v(\delta W_N^c) = \sum_{q=1}^{N} \varepsilon_N H(-\xi^3)\{(\delta\mathbf{r} - \delta\boldsymbol{\rho}) \cdot (\mathbf{n} \otimes \mathbf{n})(\mathbf{v}_S - \mathbf{v}_M)$$

$$- \xi^3 [\delta\boldsymbol{\rho}_{\xi^1} \cdot (\mathbf{n} \otimes \boldsymbol{\rho}_{\xi^1})(\mathbf{v}_S - \mathbf{v}_M) m^{11} + (\delta\mathbf{r} - \delta\boldsymbol{\rho}) \cdot (\boldsymbol{\rho}_{\xi^1} \otimes \mathbf{n}) \mathbf{v}_{M_{\xi^1}} m^{11}]\} \sqrt{\mathbf{r}_\eta \cdot \mathbf{r}_\eta}\, w_q$$

$$= \sum_{q=1}^{N} \varepsilon_N H(-\xi^3)\{\delta\mathbf{x}^T [A]^T \mathbf{n} \otimes \mathbf{n}[A]\mathbf{v}$$

$$- \xi^3 m^{11}(\delta\mathbf{x}^T [A_{\xi^1}]^T \mathbf{n} \otimes \boldsymbol{\rho}_{\xi^1}[A]\mathbf{v} + \delta\mathbf{x}^T [A]^T \boldsymbol{\rho}_{\xi^1} \otimes \mathbf{n}[A_{\xi^1}]\mathbf{v})\} \sqrt{\mathbf{r}_\eta \cdot \mathbf{r}_\eta}\, w_q. \qquad (7.64)$$

This results in the tangent matrix:

$$[K] = \sum_{q=1}^{N} \varepsilon_N H(-\xi^3)\{[A]^T \mathbf{n} \otimes \mathbf{n}[A] \qquad (7.65)$$

$$- \xi^3 m^{11}([A_{\xi^1}]^T \mathbf{n} \otimes \boldsymbol{\rho}_{\xi^1}[A] + [A]^T \boldsymbol{\rho}_{\xi^1} \otimes \mathbf{n}[A_{\xi^1}])\} \sqrt{\mathbf{r}_\eta \cdot \mathbf{r}_\eta}\, w_q. \qquad (7.66)$$

The residual $\{R\}$ and the tangent matrix $[K]$ have to be evaluated at each projected point, with "master" coordinates $\xi^1 = \xi^1_{CPP}$ for each integration point η_j calculated on the slave segment separately, and have to be summed up over all active integration points.

Remark 7.6.2 *The tangent matrix $[K]$ is symmetric as expected for a conservative system without friction.*

Chapter 18 in Part II contains all necessary information for the implementation, verification and further examples for the STS non-frictional contact element.

Remark 7.6.3 *For the high-order approximations of contact segments (classical high-order finite element, spline finite elements, isogeometric finite elements) it is necessary to implement all parts of the tangent matrix (including the curvature part) in order to keep the quadratic rate of convergence for the Newton method.*

Chapter 19 in Part II contains all necessary information for the implementation, verification and further examples for the STS non-frictional contact element with high-order approximations. The classical high-order finite element method, based on Lobatto shape functions and the Bezier spline approximations as the basis for isogeometric finite elements, is considered.

Remark 7.6.4 *Keeping the rotational part only (and curvature part in case of high order finite element within the inverted contact algorithm again allows us to obtain the effect of the following forces, as discussed in Remarks 7.4.6 and 7.5.2. One should keep in mind, however, that* in the case of STS algorithm the following forces are distributed only on the master side, *see Example 18.3.1, Chapter 18 in Part II.*

8

Verification with Analytical Solutions

This chapter is devoted to analytical solutions in contact mechanics. This area has become a big area of research, attracting many researchers in the last century. Many sophisticated mathematical instruments have been employed and aimed to find the solution of the partial differential equations describing equilibrium of bodies, including contact, see the overview in Johnson (1987). The motivation and the role for this development decreased tremendously with the development of computation mechanics since the end of last century. Nowadays, *only the major analytical solutions play significant role in computational contact mechanics – they are necessary for verification of computational algorithm*. For verification purposes, however, it is extremely important to understand under which assumptions these analytical solutions have been obtained. In this chapter, we overview the most commonly used solutions, such as the Hertz solution and the stamp solution, concentrating on the assumptions that are most important during verification. We also show original solutions for 1) rope-surface frictional interaction as a generalization of the Euler formula for belt friction and 2) for moving pendulum to find the percussion point.

8.1 Hertz Problem

One of the most famous results in contact mechanics involving a continuum is given by Hertz, see detail in Johnson (1987). This is a two body contact problem in which contacting surfaces can be locally represented by elliptic surfaces. This gave rise use of this solution for surfaces of revolution or cylindrical bodies; namely, for various wheels.

The assumptions that should be fulfilled include: the size of contact area denoted by a, the radii of curvature R_1, R_2 and the characteristic size of the body (e.g., the radius of the cylinder R in the case of a cylinder).

Introduction to Computational Contact Mechanics: A Geometrical Approach, First Edition.
Alexander Konyukhov and Ridvan Izi.
© 2015 John Wiley & Sons, Ltd. Published 2015 by John Wiley & Sons, Ltd.
Companion Website: www.wiley.com/go/Konyukhov

Thus, these assumptions follow:

1. Contact surfaces are $C2$-smooth and convex.
2. Contact surfaces are non-conforming at the contact area. This means that two surfaces contact by opposite convex parts, as shown in Figure 8.1. Inclusion of convex bodies is not allowed.
3. A contact area is much smaller than the characteristic size of the body.
4. Contact surface is slightly curved – a contact area is much smaller than the radius of curvature ($a \ll \min R_1, R_2$).
5. Stress-strain state is subjected to the linear elasticity theory (both geometrically and materially linear).
6. Contact surfaces are frictionless.

8.1.1 Contact Geometry

Taking into account the first assumption of $C2$-smoothness, we can represent each surface of a contact body described by the position vector $\rho(\xi^1, \xi^2)$ (see Section 3.3) as the Taylor expansion at the point of contact ξ_c^1, ξ_c^2:

$$\rho(\xi^1, \xi^2) = \rho(\xi_c^1, \xi_c^2) + \frac{\partial \rho}{\partial \xi^i}\Big|_{\xi_c^1, \xi_c^2} d\xi^i + \frac{1}{2}\frac{\partial^2 \rho}{\partial \xi^i \partial \xi^j}\Big|_{\xi_c^1, \xi_c^2} d\xi^i d\xi^j + ... \quad (8.1)$$

Let us introduce a global Cartesian coordinate system x, y, z at the point of contact of two bodies (see Figure 8.1) so the normal vector $\mathbf{n}$ at the contact coincides with the z-axis. Then, each surface can be locally represented (after projection equation (8.1) onto $\mathbf{n}$) in this Cartesian coordinate system as

$$z(x, y) = \frac{1}{2}\underbrace{\left(\frac{\partial^2 \rho}{\partial x^i \partial x^j} \cdot \mathbf{n}\right)}_{h_{ij}}. \quad (8.2)$$

The resulting curvature tensor h_{ij} (see equation (3.46)) can be represented with its principle values using the radius of curvatures $R_1 > 0$ and $R_2 > 0$ for the surface. Both radii are positive due to the first assumption of convexity in No. 1. Therefore, a coordinate system $x_i, y_i, z, i = 1, 2$ (rotated along z-axis) exists in which the curvature tensor is represented with its principle values as

$$[h_{ij}] : \begin{bmatrix} \frac{1}{R_1} & 0 \\ 0 & \frac{1}{R_2} \end{bmatrix}, \quad (8.3)$$

and each surface is described in the same principle directions as

$$z_i(x_i, y_i) = \frac{1}{2R_1^{(i)}}x_i^2 + \frac{1}{2R_2^{(i)}}y_i^2. \quad (8.4)$$

Verification with Analytical Solutions

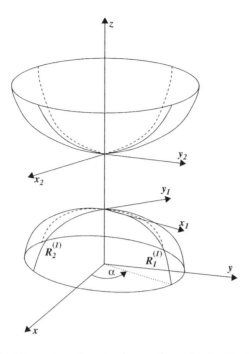

Figure 8.1 Geometry of contacting surfaces for the Hertz problem

Now, a gap function $h(x,y)$ (compare with the penetration ξ^3 in Section 4.1) is introduced in the global coordinate system x, y, z as

$$h(x,y) = z_1 - z_2. \tag{8.5}$$

Now in order to deal with them in a global reference coordinate system x, y, z, the following coordinate transformations are assumed (each coordinate system is rotated with the angle α_i along the z-axis)

$$\begin{cases} x_i = x \cos \alpha_i + y \sin \alpha_i \\ y_i = -x \sin \alpha_i + y \cos \alpha_i \end{cases} \tag{8.6}$$

Thus, the gap function in equation (8.5) is transformed as

$$h(x,y) = \frac{1}{2R_1^{(1)}} x_1^2 + \frac{1}{2R_2^{(1)}} y_1^2 - \left[-\left(\frac{1}{2R_1^{(2)}} x_2^2 + \frac{1}{2R_2^{(2)}} y_2^2 \right) \right] =$$

$$= \frac{1}{2R_1^{(1)}} (x \cos \alpha_1 + y \sin \alpha_1)^2 + \frac{1}{2R_2^{(1)}} (-x \sin \alpha_1 + y \cos \alpha_1)^2$$

$$+ \frac{1}{2R_1^{(2)}} (x \cos \alpha_2 + y \sin \alpha_2)^2 + \frac{1}{2R_2^{(2)}} (-x \sin \alpha_2 + y \cos \alpha_2)^2.$$

The second assumption of non-conforming contact surfaces that this gap function is positively determined, and therefore can be transformed in its own principle directions x' and y' as

$$h(x,y) = \frac{1}{2R'_1}x'^2 + \frac{1}{2R'_2}y'^2 \qquad (8.7)$$

where $R'_1 > 0$ and $R'_2 > 0$ are positive relative radii of curvature.

The gap function describes a deformation when a normal force P is applied, see Figure 8.2. The displacement sum of each body along the normal to the $x - y$-plane can be expressed as

$$u_{z_1} + u_{z_2} = \delta_1 + \delta_2 - h. \qquad (8.8)$$

δ_1 and δ_2 are the applied displacements parallel to the z-axis leading to the overlap as shown by the dotted lines in Figure 8.2.

The non-penetration condition is expressed as

$$u_{z_1} + u_{z_2} > \delta_1 + \delta_2 - h. \qquad (8.9)$$

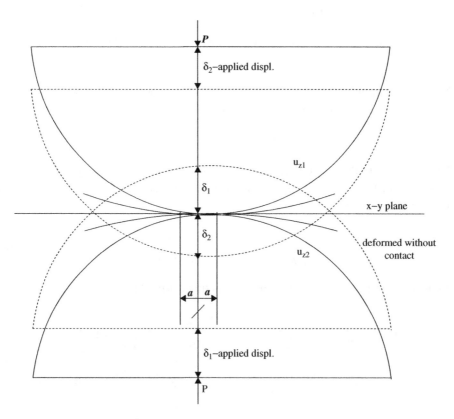

Figure 8.2 Geometry of contact and non-penetration condition for the Hertz problem

By formulating the non-penetration condition in the direction of z-axis, we *implicitly exploited assumption No. 5 of small deformations for linear elasticity (z-axis is a normal direction)*.

The next steps include the following operations:

- Consider that, for each contacting point, the Boussinesq solution for the single force acting on a half space is valid (assumptions 3, 4, and 5 are taken into account, two bodies are assumed as half spaces);
- Integrate the Boussinesq solution over the small contact area obtaining the global force P (assumptions 4 and 6). The distribution of the contact pressure over the contact area satisfying equation (8.8) should lead to the global force P.

These steps involve complicated mathematical transformations and are dropped here, the interested reader is encourage to read Johnson (1987). We represent here only cases with particular geometry.

8.1.2 Contact Pressure and Displacement for Spheres: 3D Hertz Solution

For the simplest case of two spheres, we have $R_1^{(1)} = R_2^{(1)} = R_1$; $R_1^{(2)} = R_2^{(2)} = R_2$. The vertical displacement is

$$u_{z_1} + u_{z_2} = \delta_1 + \delta_2 - \frac{1}{2}\left(\frac{1}{R_1} + \frac{1}{R_2}\right)(x^2 + y^2)$$

$$= \delta_1 + \delta_2 - \frac{1}{2R}r^2$$

where R is the relative radius of curvature; and r is the radial distance of circular contact area.

The solution result is summarized as follows:

- the contact pressure

$$p(r) = p_0\sqrt{1 - \left(\frac{r}{a}\right)^2} \quad \text{with} \quad p_0 = \frac{3P}{2\pi a^2}$$

- the radius of the contact circle

$$a = \frac{\pi p_0 R}{2E^*} = \sqrt[3]{\frac{3PR}{4E^*}}$$

- the mutual displacement of distant points

$$\delta = \frac{a^2}{R} = \sqrt[3]{\frac{9P^2}{16RE^{*2}}}.$$

where
$$\frac{1}{E^*} = \frac{1-\nu_1^2}{E_1} + \frac{1-\nu_2^2}{E_2}$$
$$\frac{1}{R} = \frac{1}{R_1} + \frac{1}{R_2}$$

8.1.3 Contact Pressure and Displacement for Cylinders: 2D Hertz Solution

The 2D solution result is summarized as follows:

- contact pressure

$$p(r) = p_0\sqrt{1 - \left(\frac{x}{a}\right)^2} \quad \text{with} \quad p_0 = \frac{2P}{\pi a}$$

the contact radius

$$a = \sqrt{\frac{4PR}{\pi E^*}}.$$

the mutual displacement

$$\delta = \frac{a^2}{2R}\left[2\ln\left(\frac{4R}{a}\right) - 1\right].$$

Remark 8.1.1 *The Hertz solution is used further for verification of the Node-To-Segment algorithm in Section 16.2.4 in Chapter 16, Part II. This verification has become the standard test to check the stress distribution for many other algorithms (Segment-To-Analytical Segment, Segment-To-Segment, etc.) especially for finite element methods with a high order of approximations, including the iso-geometric approach.*

8.2 Rigid Flat Punch Problem

A rigid punch pressed in an elastic half-space, as shown in Figure 8.3, is considered. The punch has a flat base of width $2a$, has sharp square corners and is loaded with a uniformly distributed force P ($P = 2aq$); the plane-strain condition is assumed. Since the punch is rigid, the contact area of the elastic foundation must remain flat at all contact points with the punch. With no tilt of the punch, the interface remains parallel to the undeformed surface of the solid and possesses a constant displacement normal to the interface. Various cases regarding the interaction between the punch and the elastic foundation can be considered. These are:

- contact without friction;
- full stick;
- full sliding.

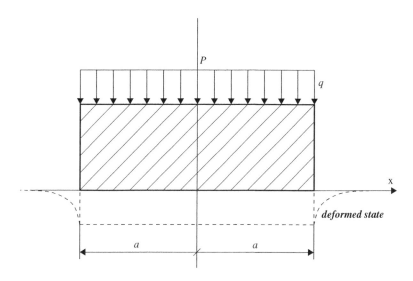

Figure 8.3 Rigid punch (stamping) problem

The solution has been constructed with the help of the Theory of Functions of a Complex Variable. For each case the pressure distribution within the contact interface in normal and tangential directions can be determined, for more details see Johnson (1987).

- For contact without friction the tangential stress distribution is $q(x) = 0$, while the normal stress is

$$p(x) = \frac{P}{\pi\sqrt{a^2 - x^2}}. \tag{8.10}$$

- For the case of *full stick* the stress distributions are given as

$$p(x) = +\frac{(1 - 2\nu)}{2(1 - \nu)} \frac{P}{\sqrt{a^2 - x^2}}$$

$$q(x) = \frac{1 - 2\nu}{2\pi^2(1 - \nu)} \frac{P}{\sqrt{a^2 - x^2}} \ln\left(\frac{a + x}{a - x}\right).$$

- For the case of the *full sliding* (quasi-statical, meaning without the dynamic forces), according the Coulomb dry friction law, the stress distributions are given as

$$p(x) = \frac{P \cos \pi\gamma}{\pi\sqrt{a^2 - x^2}} \left(\frac{a + x}{a - x}\right)^{\gamma}$$

$$q(x) = \mu_d p(x)$$

where $p(x)$ is limited by the pressure distribution recovered in equation (8.10) and γ is defined by

$$\cot(\pi\gamma) = -\frac{2(1 - \nu)}{\mu_d(1 - 2\nu)}.$$

The friction coefficient is hereby represented by μ_d.

8.3 Impact on Moving Pendulum: Center of Percussion

This example requires the simplified model of impact analytical mechanics, namely the Lagrange equations of motion for a system with an impact. We consider a moving physical pendulum (see Figure 8.4) consisting of a homogeneous rigid bar **OA** with a length l and a mass m, connected with a mass less slider at point **O**. The slider moves without friction along the x-axis. This mechanical system is characterized with two degrees of freedom: x is the position of the slider and φ is the angle. At the initial configuration with $x = 0$, $\varphi = 0$, an impact with the horizontal impulse S_X occurs at point **S** (**OS**= s). In order to study the motion of this system, we employ the Lagrange equations of motion for the system with an impact:

$$\Delta \frac{\partial T}{\partial \dot{q}_i} = S_i, i = 1,...,n. \tag{8.11}$$

Consider an impact point S at the bar:

$$\begin{cases} x_S = x + s \sin \varphi \\ y_S = s \cos \varphi \end{cases} \tag{8.12}$$

The velocity of the center of mass **C** with **OC** $= c$ is computed as

$$\begin{cases} \dot{x}_c = \dot{x} + c \cos \varphi \cdot \dot{\varphi} \\ \dot{y}_c = -c \sin \varphi \cdot \dot{\varphi} \end{cases}, \text{ with } c = \frac{l}{2}. \tag{8.13}$$

$$v_c^2 = \dot{x}^2 + (c\dot{\varphi})^2 + 2\dot{x}c\dot{\varphi} \cos \varphi \tag{8.14}$$

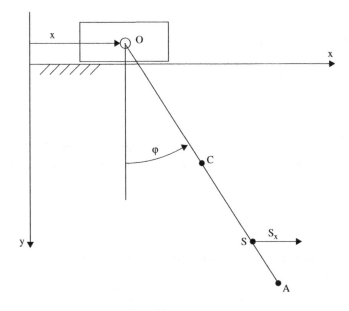

Figure 8.4 Impact on a moving pendulum – definition of the center of percussion

Verification with Analytical Solutions

Kinetic energy of the system is

$$T = \frac{J_c \dot\varphi^2}{2} + \frac{m(\dot x^2 + c^2\dot\varphi^2 + 2\dot x c\dot\varphi\cos\varphi)}{2}, \qquad (8.15)$$

where J_c is the moment of inertia around axis going through point **O**. Using the Steiner theorem $J_0 = J_c + mc^2$, the kinetic energy is written as

$$T = \frac{J_0 \dot\varphi^2}{2} + \frac{m(\dot x^2 + 2\dot x c\dot\varphi\cos\varphi)}{2}, \qquad (8.16)$$

where J_0 is the moment of inertia around the axis going through the center of mass, **C**.

The virtual work of the impulse $\mathbf{S}_x$ on the virtual displacement $\delta \mathbf{r}_S$ gives us the generalized impulses for the x and φ coordinates:

$$\delta A = (\mathbf{S}_x \cdot \delta \mathbf{r}_S) = S_x(\delta x + s\cos\varphi\,\delta\varphi). \qquad (8.17)$$

The set of necessary derivatives for the Lagrange equations (8.11) is

$$\begin{cases} \dfrac{\partial T}{\partial \dot\varphi} = J_0 \dot\varphi + m\dot x c \cos\varphi \\[1ex] \dfrac{\partial T}{\partial \dot x} = m(\dot x + c\dot\varphi \cos\varphi) \end{cases}, \qquad (8.18)$$

and the Lagrange equations (8.11) at the moment of impact $\varphi = 0$ are written as increments of the following generalized velocities:

$$\begin{cases} \Delta(m\dot x + mc\dot\varphi) = S_x \\ \Delta(J_0\dot\varphi + m\dot x) = sS_x \end{cases}, \qquad (8.19)$$

And taking into account the initial conditions:

$$\text{at } \varphi = 0 \Rightarrow \dot\varphi_0 = 0 \text{ and } \dot x_0 = 0 \qquad (8.20)$$

the system finally becomes

$$\begin{cases} m\dot x + mc\dot\varphi = S_x \\ m\dot x c + J_0\dot\varphi = sS_x \end{cases}. \qquad (8.21)$$

The determinant of the system is proportional to the central moment of inertia J_c and is not zero

$$\det = mJ_0 - m^2c^2 = m(J_0 - mc^2) = mJ_c \neq 0. \qquad (8.22)$$

The solution of the system is

$$\begin{cases} \dot\varphi = \dfrac{S_x}{m}(s - x) \\[1ex] \dot x = \dfrac{S_x}{m}(J_0 - mcs) \end{cases}. \qquad (8.23)$$

One can observe that the motion of the slider depends on the position of the impact s. The slider moves to the right if $J_0 - mcs > 0$; and it moves to the left if $J_0 - mcs < 0$.

Definition 8.3.1 *The point of impact S, determined by the condition $J_0 - mcs = 0$, is called a center of percussion.*

One can observe that the impact at the center of mass $s = c$ causes only the translational motion without angular velocity $\dot{\varphi} = 0$.

The center of percussion for the homogeneous bar is determined as

$$s = \frac{J_0}{mc} = \frac{ml^2/3}{l/2} = \frac{2}{3}l. \qquad (8.24)$$

The point of percussion has the following characteristics:

- After the horizontal impact at the center of percussion of the moving pendulum, the slider is not moving.
- After the horizontal impact at the center of percussion of the pendulum (slider is fixed), the reaction impulse at the pivot point (fixed slider) is zero.

The last fact has found a vast practical application for various impact instruments.

Remark 8.3.2 *This example is used to verify impact problems in dynamics, see Section 24.2.3 in Chapter 24, Part II.*

8.4 Generalized Euler–Eytelwein Problem

The famous Euler–Eytelwein problem for the equilibrium of rope on a rough cylinder under pulling forces has the solution $T = T_0 e^{\mu\varphi}$, where μ is the coefficient of friction between the rope and the cylinder, and $\Delta\varphi = \varphi - \varphi_0$ is a polar angle spanning the rope. This solution has been recently generalized for the case of contact between an arbitrary curved rope on an arbitrary curved orthotropic surface in Konyukhov (2013). Here, we represent a simplified for a 2D geometry result.

Theorem 8.4.1 *If a rope lays in equilibrium under tangential forces on a rough curve in 2D (see Figure 8.5) then the following conditions (all of them) are satisfied:*

1. *No separation – normal reaction N is positive for all points of the curve:*

$$N = -kT = -\frac{d\varphi}{ds}T > 0 \qquad (8.25)$$

This condition can be interpreted as:
- *curvature k is keeping its sign; or*
- *angle φ is a monotone function.*

Verification with Analytical Solutions

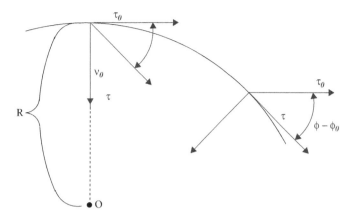

Figure 8.5 Plane 2D curve: Interpretation of the curvature $k = \frac{1}{R} = \frac{d\varphi}{ds}$

2. *Limit values of the tangential forces:*
   The forces at both ends T and T_0 satisfy the following inequality

$$T_0 e^{-\mu_\tau(\varphi_0-\varphi)} < T < e^{\mu_\tau(\varphi_0-\varphi)} \tag{8.26}$$

where the μ_τ is coefficient of friction between a rope and a curve in the tangential direction τ (the coefficient of friction in the direction orthogonal to the Figure 8.5 should be positive); and $\varphi_0 - \varphi$ is the final angle of rotation for the unit tangent τ, or the increment of the slope of the tangent line to the graph $\hat{\alpha} - \hat{\alpha}_0$ (see Figure 8.5).

It is remarkable that the form of the famous Euler solution $\frac{T}{T_0} = e^{\mu_\tau(\varphi_0-\varphi)}$ is completely preserved. *The result gives a rise to a very elegant geometrical solution to the problem.*

8.4.1 A Rope on a Circle and a Rope on an Ellipse

Consider a 2D problem of a rope laying on an ellipse and compare this solution with the classical Euler–Eytelwein statement – a rope on a circle (see Figure 8.6). For the ellipse, defined in the polar coordinates as

$$x = a\cos\varphi;\, y = b\sin\varphi, \tag{8.27}$$

the slope of the tangent line is calculated as

$$\tan\hat{\alpha} = \frac{y'}{x'} = -\frac{b\cos\varphi}{a\sin\varphi}, \tag{8.28}$$

which can be directly used in formula (8.26). The simple geometrical interpretation gives us the same ratio of the forces with the same angle $\varphi - \varphi_0$, for example in both

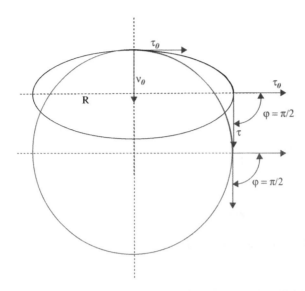

Figure 8.6 Comparison between a rope wrapped over a circular cylinder and a rope on an ellipse in 2D: The same ratio of forces for the same angle $\Delta\varphi = \frac{\pi}{2}: \rightarrow \frac{T}{T_0} = e^{\mu_\tau \frac{\pi}{2}}$

cases of a rope wrapped over a quarter of a circle and a rope wrapped over a quarter of an ellipse: $\frac{T}{T_0} = e^{\mu_\tau \frac{\pi}{2}}$, see Figure 8.6.

Remark 8.4.2 *The last result is used for verification of computational algorithms, including frictional contact, in Section 22.2.4 in Chapter 22, Part II.*

9

Frictional Contact Problems

In previous chapters, we were concentrating on various types of discretization for only non-frictional problems. In order to consider frictional interaction, the frictional tangential part in equation (6.7) should be taken into account. Additive inclusion allows us just to consider the frictional part independent of the normal part. This chapter is devoted to the description of the frictional tangential part only; therefore, the pure tangential part δW_T^c is taken from the weak form in equation (6.7) and is expressed in the local coordinate system as

$$\delta W_T^c = \int_S T^1(\rho_1 \cdot \rho_1) \delta \xi^1 \, ds. \tag{9.1}$$

In order to deal with the frictional interaction a constitutive relationship should be supplied for the tangential component T^1: sticking and sliding cases should be defined. This chapter deals with the most famous law of frictional interaction – the Coulomb friction law. The regularization technique, return-mapping algorithm and further linearization are constructed in a covariant form for further computational algorithms, leading again to various contact elements (NTN, NTS with linear and high order approximations), depending on the discretization techniques studied in Chapter 7.

9.1 Measures of Contact Interactions – Sticking and Sliding Case: Friction Law

Various situations in kinematics of the tangential interaction are distinguished with the help of measuring tangential interaction (see Section 4.1). If the measure of normal contact interaction is taken in the form of the coordinate ξ^3 as penetration, then the measure of tangential interaction is taken in the rate form as $\dot{\xi}^i$, see Figure 9.1.

Introduction to Computational Contact Mechanics: A Geometrical Approach, First Edition.
Alexander Konyukhov and Ridvan Izi.
© 2015 John Wiley & Sons, Ltd. Published 2015 by John Wiley & Sons, Ltd.
Companion Website: www.wiley.com/go/Konyukhov

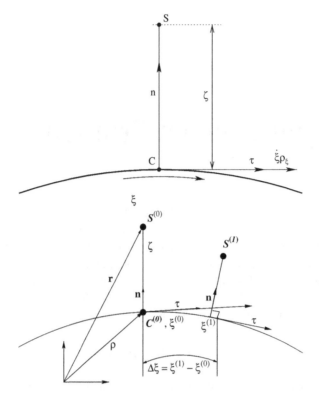

Figure 9.1 Kinematics of frictional contact: Computation of the tangential measure $\Delta\xi$

The full contact traction vector is given as

$$\mathbf{R} = N\mathbf{n} + T^1 \boldsymbol{\rho}_1, \tag{9.2}$$

where the tangential force vector $\mathbf{T} = T^1 \boldsymbol{\rho}_1$ is written in contravariant components.

Definition 9.1.1 *Considering the relative velocity $\mathbf{v}_S - \mathbf{v}_M$ in the tangential direction $\boldsymbol{\rho}_1$ (see equation (4.24)) as convective tangential velocity $\dot{\xi}^1$, two stages – sticking and sliding – are distinguished:*

- Stage "sticking"
  Relative velocity is zero $\dot{\xi}^1 = 0$ on the tangent plane $\xi^3 = 0$ and tangential force vector $T^1 \boldsymbol{\rho}_1$ is obtained from the global equilibrium equation.
- Stage "sliding"
  Relative velocity is not zero $\dot{\xi}^1 \neq 0$ and an additional constitutive relation is supplied in order to define the sliding force.

9.1.1 Coulomb Friction Law

The most famous constitutive relationship between sliding force and relative sliding velocity is the Coulomb friction law, specified for sticking and sliding cases as follows:

$$\begin{cases} \text{sticking: } \dot{\xi}^1 = 0, & |\mathbf{T}| \leq \mu_s |\mathbf{N}|, \quad \mathbf{T} \text{ is derived from equilibrium equation} \\ \text{sliding: } \dot{\xi}^1 \neq 0, & |\mathbf{T}| > \mu_s |\mathbf{N}|, \quad \mathbf{T} = -\dfrac{\dot{\xi}^1}{\left|\dot{\xi}^1\right|} \mu_d |\mathbf{N}| \end{cases} \tag{9.3}$$

where μ_s is the sticking coefficient and μ_d the sliding coefficient of friction. In the simplest model we have simply the coefficient of friction $\mu_s = \mu_d = \mu$, while the complex model may include dependence of μ_d on the relative sliding velocity.

The absolute value of the normal force is computed as $|\mathbf{N}| = N$, and the tangential force vector (see equation (6.5)) is given as $\mathbf{T} = T^1 \boldsymbol{\rho}_1$ with the absolute value as $|\mathbf{T}| = |T^1| \|\boldsymbol{\rho}_1\|$.

As stated already in Section 6.2, the contact vector components N and T^1 are additional unknowns in the contact integral equation (6.7). Regularization is necessary in order to compute the tangential traction T^1. This regularization is based on an elasto-plastic analogy to model the Coulomb dry friction law leading to penalty regularization and the return-mapping scheme.

9.2 Regularization of Tangential Force and Return Mapping Algorithm

The regularization of the tangential force assumes that the rate of the tangent force is proportional to the relative velocity $\mathbf{v}_r = \mathbf{v}_M - \mathbf{v}_s$ on the tangent plane – in this case elastic tangential deformations are allowed

$$\frac{d\mathbf{T}}{dt} = -\varepsilon_T \mathbf{v}_r, \tag{9.4}$$

where covariant derivative is involved, see equation (3.68) in Section 3.4.3. In the case of 3D contact, equation (9.4) is written for contravariant components T^i as

$$\frac{\partial T^i}{\partial t} + \left(\frac{\partial T^i}{\partial \xi^j} + \Gamma^i_{kj} T^k \right) \dot{\xi}^j + h^i_j \dot{\xi}^3 T^j = -\varepsilon_T \dot{\xi}^i, \tag{9.5}$$

where Γ^k_{ij} are Christoffel symbols for the surface, see equation (3.63).

Considering a 2D case as a simplified cylindrical case of 3D geometry we obtain

$$\frac{dT^1}{dt} = -\varepsilon_T \dot{\xi}^1 \tag{9.6}$$

where ε_T represents the tangential penalty parameter.

Remark 9.2.1 *Equations are valid for both dynamic and quasi-static computations. The time t is a real time for dynamic problems and just a loading parameter for quasi-static computations.*

Assuming now quasi-static computations, the finite difference scheme with a load increment of $\Delta t = 1$ is then applied to equation (9.6)

$$T^1_{(n+1)} - T^1_{(n)} = -\varepsilon_T(\xi^1_{(n+1)} - \xi^1_{(n)}) = -\varepsilon_T \Delta \xi^1 \qquad (9.7)$$

where the indexes (n) and $(n+1)$ refer to consecutive load steps. The solution of ordinary differential equation (9.6) requires initial conditions. Usually, the unstressed configuration at the initial sticking point $\xi^1_{(0)}$ is assumed, thus leading to the following conditions

$$T^1_{(0)} = 0; \quad \xi^1 = \xi^1_{(0)}, \qquad (9.8)$$

where the unstressed configuration $T^1_{(0)} = 0$ is set at the "initial sticking point" $\xi^1_{(0)}$, that, for example, can be obtained as a result of the CPP procedure for the bodies without loading. One can clearly see that the use of this finite difference scheme requires the storage of the variables $T^1_{(n)}$ and $\xi^1_{(n)}$ of the previous load step (n) for implementation. Continuing the calculation recursively in equation (9.7) we can obtain

$$\begin{aligned}
T^1_{(n+1)} &= T^1_{(n)} - \varepsilon_T(\xi^1_{(n+1)} - \xi^1_{(n)}) = \\
&= T^1_{(n-1)} - \varepsilon_T(\xi^1_{(n)} - \xi^1_{(n-1)}) - \varepsilon_T(\xi^1_{(n+1)} - \xi^1_{(n)}) = \\
&= T^1_{(n-1)} - \varepsilon_T(\xi^1_{(n+1)} - \xi^1_{(n-1)}) = \cdots = \\
&= T^1_{(0)} - \varepsilon_T(\xi^1_{(n+1)} - \xi^1_{(0)}).
\end{aligned} \qquad (9.9)$$

If $T^1_{(n+1)}$ is computed with regard to the unstressed configuration with an initial sticking point $\xi^1_{(0)}$, then the tangential traction can be computed as

$$T^1_{(n+1)} = -\varepsilon_T(\xi^1_{(n+1)} - \xi^1_{(0)}) = -\varepsilon_T \Delta \xi^1. \qquad (9.10)$$

In this case only the initial sticking point $\xi^1_{(0)}$ is stored.

Definition 9.2.2 *The return-mapping scheme is a general name for the numerical scheme in computational mechanics for computing the real parameter based on the following steps:*

- *computation of the trial parameter based on the assumed behavior (usually elastic), and*
- *final correction of the real parameter based on the constitutive law for the yield condition.*

Frictional Contact Problems

The return-mapping scheme for the Coulomb frictional law in equation (9.3) is formulated as follows:

- *The trial step.*
  The trial tangential force is assumed to be elastic, $T^{1,trial}_{(n+1)} = -\varepsilon_T \Delta \xi^1$, see equation (9.10).
- *The return mapping (step).*
  The contravariant component of the real tangential force is computed with regard to the Coulomb friction law

$$T^1 = \begin{cases} T^{1,\,trial}_{(n+1)} & \text{if } \left|T^{1,\,trial}_{(n+1)}\right| \leq \dfrac{\mu_S |N|}{|\boldsymbol{\rho}_1|} \text{ sticking} \\[2ex] -\dfrac{\Delta \xi^1}{|\Delta \xi^1|} \dfrac{\mu_d |N|}{|\boldsymbol{\rho}_1|} & \text{if } \left|T^{1,\,trial}_{(n+1)}\right| > \dfrac{\mu_S |N|}{|\boldsymbol{\rho}_1|} \text{ sliding.} \end{cases} \quad (9.11)$$

If the penalty regularization for the normal traction N is taken into account, see equation (6.9) in Section 6.2, then $|N| = \varepsilon_N |\xi^3|$.

9.2.1 Elasto-Plastic Analogy: Principle of Maximum of Dissipation

The return-mapping algorithm in equation (9.11) can be derived from the more general principle used in thermodynamics – the principle of maximum dissipation. The frictional problem can be stated as follows (elasto-plastic analogy):

1. The whole tangential relative velocity/displacement is split additively into elastic and inelastic (sliding) parts.
   In the rate form: $\mathbf{v}^r = \mathbf{v}^r_{(el)} + \mathbf{v}^r_{(sl)}$
   In the incremental form: $\Delta \xi^i = \Delta \xi^i_{el} + \Delta \xi^i_{sl}$.
2. The elastic tangential force is subjected to the following elastic constitutive relationship:
   In the rate form:
   $$\frac{d\mathbf{T}}{dt} = -\varepsilon_T \mathbf{v}^r_{el}$$
   In the incremental form for contravariant component:
   $$T^1_{tr}{}^{(n+1)} = T^1{}^{(n)} - \varepsilon_T \Delta \xi^1.$$
3. Tangential force is subjected to the constitutive relation determining the region of inelastic deformation (yield function):
   $$\Phi = |\mathbf{T}| - \mu |N|$$
   if $\Phi < 0$ then force $\mathbf{T}$ is elastic
   if $\Phi \geq 0$ then force $\mathbf{T}$ is inelastic (sliding).

The dissipation function is introduced as:

$$D = \Delta \xi_{sl} \cdot \mathbf{T}_{sl} = \Delta \xi_{sl}^1 T_1^{sl}. \tag{9.12}$$

In order to determine the inelastic (sliding) force and further parameters, the *principle of maximum of dissipation* is formulated as follows:

The sliding force and further parameters are defined as the solution of the maximization problem for the dissipation function in equation (9.12) subjected to conditions 1), 2), 3) (the constraint optimization problem).

The optimization problem is solved via the Lagrange multiplier method. We formulate the minimization problem for the following Lagrange function for each step $(n+1)$:

$$L(\Delta \xi^{sl}, \lambda) := -\mathbf{T}^{sl} \cdot \Delta \xi^{sl} + \lambda(|\mathbf{T}^{tr}| - \mu|N|) \to \min \tag{9.13}$$

with the Karush–Kuhn–Tucker conditions:

1. $\Phi < 0, \lambda \geq 0$ sticking
2. $\Phi \geq 0, \lambda > 0$ sliding
3. $\lambda \Phi = 0$.

Employing the necessary condition for a minimum via the contravariant component $\Delta \xi_{sl}^1$

$$\frac{\partial L}{\partial \Delta \xi_{sl}^1} = 0, \tag{9.14}$$

we obtain the covariant component of the sliding force as

$$T_1^{sl} = \lambda \frac{\partial}{\partial \Delta \xi_{sl}^1} \left[|\mathbf{T}^{tr}| - \mu|N| \right]$$

$$= \lambda \frac{\partial}{\partial T_{tr}^1} \left[\sqrt{T_{tr}^1 T_{tr}^1 m_{11}} \right] \frac{\partial T_{tr}^1}{\partial \Delta \xi_{el}^1} \frac{\partial \Delta \xi_{el}^1}{\partial \Delta \xi_{sl}^1}$$

$$= \lambda \frac{T_{tr}^1 m_{11}}{\sqrt{T_{tr}^1 T_{tr}^1 m_{11}}} (-\varepsilon_T)(-1) = \lambda \varepsilon_T \frac{T_1^{tr}}{|\mathbf{T}^{tr}|}. \tag{9.15}$$

Here, we carefully consider co- and contravariant components and their transformation with a metric coefficient m_{11} in 2D. Using now the yield condition $\Phi = 0$ for sliding

$$\phi := \sqrt{\mathbf{T}^{sl} \cdot \mathbf{T}^{sl}} - \mu|N| = \sqrt{\lambda^2 \varepsilon_T^2 \frac{T_1^{tr} T_1^{tr} m^{11}}{|\mathbf{T}^{tr}|^2}} - \mu|N| = 0, \tag{9.16}$$

the positive Lagrange multiplier is determined as

$$\lambda \varepsilon_T = \mu|N|. \tag{9.17}$$

Frictional Contact Problems

Now we can derive the contravariant component in equation (9.15) as

$$T^1_{sl} = T^{sl}_1 m^{11} = \mu |N| \frac{T^1_{tr}}{|\mathbf{T}^{tr}|} \qquad (9.18)$$

and expressing T^1_{tr} via the increments $\Delta \xi^1$ as

$$T^1_{sl} = T^{sl}_1 m^{11} = \mu |N| \frac{-\varepsilon_T \Delta \xi^1}{\sqrt{\varepsilon_T \Delta \xi^1 \varepsilon_T \Delta \xi^1 m_{11}}} = -\frac{\Delta \xi^1}{|\Delta \xi^1|} \frac{\mu |N|}{\sqrt{m_{11}}}, \qquad (9.19)$$

which coincides exactly with the sliding force in equation (9.11).

9.2.2 Update of Sliding Displacements in the Case of Reversible Loading

We consider a geometrical interpretation of the return-mapping scheme in equation (9.11) together with the evolution equation (9.10) for the trial tangential force, see Figure 9.2.

$$|T^1_{tr,\,(m)}| < \mu \frac{|N^{(m)}|}{\sqrt{m_{11}}} \quad \Rightarrow \quad \varepsilon_T |\xi^{(m)} - \xi^0| < \mu \frac{|N^{(m)}|}{\sqrt{m_{11}}} \qquad (9.20a)$$

$$|\xi^{(m)} - \xi^0| < R^{(m)}_\xi, \quad R^{(m)}_\xi = \frac{\mu |N^{(m)}|}{\varepsilon_T \sqrt{m_{11}}} \qquad (9.20b)$$

Equation (9.20b) describes an allowable elastic region $\mathbf{A}^{(m)}\mathbf{B}^{(m)}$ with a center of attraction, $\mathbf{O}^{(m)}$ (initial sticking point). All points inside this domain are in the "sticking condition". If now a point $\xi^{(m+1)}$ appears to be outside of the domain at load step $(m+1)$, then its only admissible position is on the boundary of the domain, i.e. must

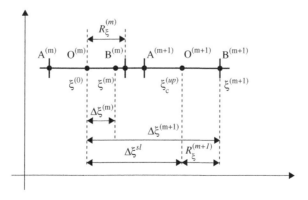

Figure 9.2 Coulomb friction. Updating of sliding displacements in convective coordinates. Motion of the elastic region and update of the center as initial sticking point

coincide with $\mathbf{B}^{(m+1)}$. A sliding force is applied then at the contact point, see equation (9.11). As long as we have a motion of the contact point only in one direction the sign function for the sliding force $sgn(T_{tr}^{(m+1)}) = sgn(\Delta\xi^{(m+1)})$ does not change and the computation will be correct. However, when a reversible load is applied that forces the contact point to move forward or backward, the attraction point $\mathbf{O}^{(m)}$ must be updated in order to define the sign function for the sliding force correctly. This update can be defined geometrically from Figure 9.2:

$$|\Delta\xi^{(m+1)}| = |\Delta\xi_{sl}| + R_\xi^{(m+1)} = |\Delta\xi^{(m+1)}| - \frac{\mu|N^{(m)}|}{\varepsilon_T}. \qquad (9.21)$$

The absolute value of the sliding displacement is then computed at load step $(m+1)$ as:

$$|\Delta\xi_{sl}| = |\xi^{(m+1)} - \xi^{(0)}| - \frac{\mu|N^{(m+1)}|}{\varepsilon_T}, \qquad (9.22)$$

and the updated center of the elastic domain becomes:

$$\xi_c^{(up)} = \xi^{(0)} + sgn(\xi^{(m+1)} - \xi^{(0)})|\Delta\xi_{sl}|. \qquad (9.23)$$

For the next step, the evolution equation (9.10) is corrected as

$$T^{1\,(m+2)} = -\varepsilon_T(\xi^{(m+2)} - \xi_c^{up}). \qquad (9.24)$$

Remark 9.2.3 *As an alternative procedure the back-substitution of the evolution equation (9.10) into equation (9.22) gives the updated scheme via the trial force:*

$$|\Delta\xi_{sl}| = \frac{1}{\varepsilon_T}\left(|T_{tr}^{(m+1)}| - \mu|N^{(m+1)}|\right). \qquad (9.25)$$

Remark 9.2.4 *In general 3D case the initial sticking point $\xi^{(0)}$ is updated on the tangent plane after the detection of the sliding at the point $\boldsymbol{\xi}^{(m+1)}$ as follows:*

$$\boldsymbol{\xi}_m^{(0)} = \boldsymbol{\xi}^{(m+1)} - \frac{(\boldsymbol{\xi}^{(m+1)} - \boldsymbol{\xi}^{(0)})}{|\boldsymbol{\xi}^{(m+1)} - \boldsymbol{\xi}^{(0)}|}\frac{\mu|N|}{\varepsilon_T}, \qquad (9.26)$$

where $\boldsymbol{\xi}_m^{(0)}$ – is the updated center of the elastic region; $\boldsymbol{\xi}^{(m+1)} - \boldsymbol{\xi}^{(0)}$ – is the whole pass including elastic and sliding (plastic) parts, $\frac{\mu|N|}{\varepsilon_T}$ – is a radius of the elastic circular region.

9.3 Weak Form and its Consistent Linearization

The weak form now contains two parts, see equation (6.7) in Section 6.1, for the normal δW_N^c and frictional interaction δW_T^c

$$\delta W^c = \int_s [\underbrace{N\delta\xi^3}_{\delta W_N^c} + \underbrace{T^1(\boldsymbol{\rho}_{\xi^1} \cdot \boldsymbol{\rho}_{\xi^1})\delta\xi^1}_{\delta W_T^c}]ds. \qquad (9.27)$$

Frictional Contact Problems

The discretization of the weak form leads then to two additive parts of the residual vector: $[R_N]$ for the normal interaction, which has been considered already in Chapter 7, and $[R_T]$ for the tangential interaction. The tangential force in the tangential part is computed with regard to the return mapping algorithm in equation (9.11). Thus, *the linearization of the weak form should be separately considered for sticking and sliding cases*. For the linearization we will use both forms:
in convective coordinates

$$\delta W_T^c = \int_s T^1 (\boldsymbol{\rho}_{\xi^1} \cdot \boldsymbol{\rho}_{\xi^1}) \delta \xi^1 ds_s; \qquad (9.28)$$

and in the Frenet coordinate system via the arc-length s

$$\delta W_T^c = \int_s T^1 \delta s ds_s, \qquad (9.29)$$

where the slave segment s_s emphasizes the integration over the slave segment.

The consistent linearization follows, in general, the following steps

$$D_v(\delta W_T^c) = \int_s D_v(T^1)(\boldsymbol{\rho}_{\xi^1} \cdot \boldsymbol{\rho}_{\xi^1}) \delta \xi^1 ds_s \qquad (9.30)$$

$$+ \int_s T^1 D_v(\boldsymbol{\rho}_{\xi^1} \cdot \boldsymbol{\rho}_{\xi^1}) \delta \xi^1 ds_s$$

$$+ \int_s T^1 (\boldsymbol{\rho}_{\xi^1} \cdot \boldsymbol{\rho}_{\xi^1}) D_v(\delta \xi^1) ds_s. \qquad (9.31)$$

Since it is tremendously simplified in the case of small displacements, these operations are considered separately for NTN and NTS approaches.

9.4 Frictional Node-To-Node (NTN) Contact Element

The frictional Node-To-Node (NTN) contact element possesses the same kinematics studied for non-frictional NTN element in Section 7.2. We are recalling here the same strict assumptions as for the small displacement problem for contact problems, which are valid during the deformation process:

1. Two contacting nodes are approximately laying on one normal vector **n**.
2. Normal vector **n** (and therefore tangent vector $\boldsymbol{\tau}$) is changing negligibly.
3. Displacements of two contacting nodes are small in comparison to the size of the finite elements used for this discretization.
4. For the discretization of both contacting bodies *only linear finite elements* are used.

These assumptions allows to specify two constant vectors:
a unit normal vector $|\mathbf{n}| = 1$:

$$\mathbf{n} = \begin{Bmatrix} n_x \\ n_y \end{Bmatrix} \qquad (9.32)$$

and a unit tangent vector $|\boldsymbol{\tau}| = 1$

$$\boldsymbol{\tau} = \begin{Bmatrix} \tau_x \\ \tau_y \end{Bmatrix} \qquad (9.33)$$

The vectors satisfy *a priori* the orthogonality condition: $\mathbf{n} \cdot \boldsymbol{\tau} = 0$. We are employing the same parameters (nodal vector, approximation matrix, etc.) necessary for discretization of the NTN approach (see Section 7.2 and Figure 13.1 in Chapter 13, Part II).

The measures of contact interaction is computed as follows:

- Measure of normal interaction – penetration (see equation (7.9))

$$\xi^3 = (\mathbf{r} - \boldsymbol{\rho}) \cdot \mathbf{n} = (\mathbf{r} - \boldsymbol{\rho})^T \mathbf{n} = \mathbf{x}^T [A]^T \mathbf{n}. \qquad (9.34)$$

- Measure of tangential interaction is derived from the convective velocity

$$\dot{\xi}^1 = (\mathbf{v}_s - \mathbf{v}_M) \cdot \boldsymbol{\tau} \qquad (9.35)$$

in which direct computation is possible due to $\boldsymbol{\tau} = constant$:

$$\Delta \xi^1 = (\mathbf{u}_s - \mathbf{u}_M) \cdot \boldsymbol{\tau} \qquad (9.36)$$

$\mathbf{u}_s$, $\mathbf{u}_M$ are displacement vectors of the slave and master nodes correspondingly. The relative displacement vector $(\mathbf{u}_s - \mathbf{u}_M)$ is discretized with the approximation matrix $[A]$ for the NTN approach in equation (7.6)

$$\mathbf{u}_s - \mathbf{u}_M = [A]\mathbf{u}. \qquad (9.37)$$

9.4.1 Regularization of the Contact Conditions

The normal force for the NTN approach is computed in the same fashion as for the NTN non-frictional approach in Section 7.2.

$$N = H(-\xi^3)\varepsilon_N \xi^3 \qquad (9.38)$$

The return-mapping in equation (9.11) is simplified due to given constant normal and tangent vector.

The trial tangential force is computed as

$$T^{trial} = -\varepsilon_T \Delta \xi^1 = \qquad (9.39)$$
$$= -\varepsilon_T (\mathbf{u}_s - \mathbf{u}_c) \cdot \boldsymbol{\tau} = -\varepsilon_T [(u_{x_s} - u_{x_M}) \cdot \tau_x + (u_{y_s} - u_{y_M}) \cdot \tau_y].$$

Here, ε_τ is the tangential penalty. The computation of the increment in the form of equation (9.36) fulfills automatically the initial condition: *no tangent force at initial undeformed configuration* with $T = 0$ with $\mathbf{u} = 0$.

Return-mapping scheme.

$$T^1 = \begin{cases} T^{1,\ trial}_{(n+1)} & \text{if } \left|T^{1,\ trial}_{(n+1)}\right| \leq \mu_S |N| \text{ sticking} \\ -sign(\Delta\xi^1)\mu_d |N| & \text{if } \left|T^{1,\ trial}_{(n+1)}\right| > \mu_S |N| \text{ sliding.} \end{cases} \quad (9.40)$$

9.4.2 Linearization the of Tangential Part for the NTN Contact Approach

The full contact integral is computed with regard to the numerical integration for the Node-To-Node (NTN) as discussed in Section 7.1.1:

$$\delta W_c = \int_l (T^1 \delta\xi^1 + N\delta\xi^3) dl = \underbrace{T^1 \delta\xi^1}_{\delta W_T^c} + \underbrace{N\delta\xi^3}_{\delta W_N^c}. \quad (9.41)$$

Due to the strict assumptions for the NTN contact approach, the linearization for the frictional NTN is tremendously simplified. Due to the assumption of small displacements and constant unit normal vector we have:

$$\frac{d}{dt}\delta\xi^1 = 0; \quad \frac{d}{dt}\delta\xi^3 = 0. \quad (9.42)$$

The tangential part δW_T^c, therefore, is linearized as follows:

- for the sticking case:

$$\frac{d\delta W_c^T}{dt} = \frac{dT^1_{trial}}{dt}\delta\xi^1 = -\varepsilon_T \dot{\xi}^1 \delta\xi^1; \quad (9.43)$$

- for the sliding case:

$$\frac{d\delta W_c^T}{dt} = \frac{dT^1_{slid}}{dt}\delta\xi^1 = \frac{d}{dt}\left(-sign(\Delta\xi^1)\mu |N|\delta\xi^1\right)$$
$$= -sign(\Delta\xi^1)\mu\varepsilon_N \dot{\xi}^3 \delta\xi^1. \quad (9.44)$$

9.4.3 Discretization of Frictional NTN

The approximation matrix in equation (7.6) introduced for the NTN approach in Section 7.2 is used further to obtain the residual and tangent matrix for tangential part.

9.4.3.1 Full Residual for Frictional NTN Contact Element

We repeat for necessity here the discretization of a normal part of the residual from equation (7.11) leading to the residual vector for the normal part

$$[R_N] = \varepsilon_N \xi^3 \{n\}^T [A] = \varepsilon_N \xi^3 \begin{Bmatrix} -n_x \\ -n_y \\ +n_x \\ +n_y \end{Bmatrix} \tag{9.45}$$

The discretization of the residual vector for the tangent part

$$\delta W_c^T = T^{real} \delta \xi^1 = T^{real} (\delta \mathbf{r}_s - \delta \boldsymbol{\rho}_c) \cdot \boldsymbol{\tau} = T^{real} \{\tau\}^T [A] \{\delta x\} \tag{9.46}$$

leads to the following residual vector for the tangential part

$$[R_T] = T^{real} \{\tau\}^T [A] = T^{real} \begin{Bmatrix} -\tau_x \\ -\tau_y \\ \tau_x \\ \tau_y \end{Bmatrix}. \tag{9.47}$$

The full residual is derived as a sum of both parts

$$[R] = [R_N] + [R_T]. \tag{9.48}$$

9.4.3.2 Full Tangent Matrix for Frictional NTN Contact Element

Discretization of the linearized functional δW_T^c leads to the following matrices:

- Tangential part for sticking is obtained from equation (9.43)

$$\frac{d\delta W_c^T}{dt} = -\varepsilon_T \dot{\xi}^1 \delta \xi^1 = -\varepsilon_T (\mathbf{v}_s - \mathbf{v}_c) \cdot \boldsymbol{\tau} \cdot (\delta \mathbf{r}_s - \delta \boldsymbol{\rho}_c) \cdot \boldsymbol{\tau} \tag{9.49}$$

$$= -\varepsilon_T (\delta \mathbf{r}_s - \delta \boldsymbol{\rho}_c) \cdot \boldsymbol{\tau} \otimes \boldsymbol{\tau} (\mathbf{v}_s - \mathbf{v}_c) = -([A]\{\delta x\})^T \varepsilon_T \boldsymbol{\tau} \otimes \boldsymbol{\tau} [A]\{v\}$$

$$= -\{\delta x\}^T [A^T] \varepsilon_T \boldsymbol{\tau} \otimes \boldsymbol{\tau} [A]\{v\}. \tag{9.50}$$

Thus, the tangent matrix for sticking is

$$[K_T^{stick}] = -\varepsilon_T [A]^T \boldsymbol{\tau} \otimes \boldsymbol{\tau} [A]. \tag{9.51}$$

- Tangential part for sliding is obtained from equation (9.44)

$$\frac{d\delta W_c^T}{dt} = -sign(\delta \xi^1) \cdot \mu \varepsilon_N \dot{\xi}^3 \delta \xi^1 \tag{9.52}$$

$$= -sign(\Delta \xi^1) \cdot \mu \varepsilon_N (\mathbf{r}_s - \mathbf{r}_c) \cdot \mathbf{n} (\delta \mathbf{r}_s - \delta \mathbf{r}_c) \cdot \boldsymbol{\tau}$$

$$= -(\delta \mathbf{r}_s - \delta \mathbf{r}_c) \cdot sign(\delta \xi^1) \mu \varepsilon_N \boldsymbol{\tau} \otimes \mathbf{n} (\mathbf{r}_s - \mathbf{r}_c)$$

$$= -([A]\{\delta x\})^T \cdot sign(\Delta \xi^1) \cdot \mu \varepsilon_N \boldsymbol{\tau} \otimes \mathbf{n} [A]\{v\}. \tag{9.53}$$

Frictional Contact Problems

Thus, the tangent matrix for sliding is

$$[K_T^{slide}] = -sign(\Delta \xi^1) \cdot \mu \varepsilon_N [A]^T \cdot \boldsymbol{\tau} \otimes \mathbf{n}[A]. \tag{9.54}$$

The full tangent matrix contains both the normal part $[K_N]$ in equation (7.14) and corresponding tangential part $[K_T]$ depending on the result of the return-mapping scheme in equation (9.40).

$$[K] = [K_N] + [K_T]. \tag{9.55}$$

Remark 9.4.1 *The full matrix in the case of sticking is symmetric, but in the case of sliding is unsymmetric (namely due to equation (9.54)). The linear equation system solver for further numerical solution, which is usually used for symmetric matrices for structural problems, should be changed into the solver for non-symmetric matrices.*

9.4.4 Algorithm for a Local Level Frictional NTN Contact Element

We list here schematically the necessary steps for the implementation of the NTN frictional element, Chapter 21 in Part II contains all necessary information for the implementation, verification and further typical examples of the NTN approach with the Coulomb friction law based on the Penalty method.

1. Update coordinates
2. Compute penetration and check penetration

$$\xi^3 = (\mathbf{r}_s - \boldsymbol{\rho}_c) \cdot \mathbf{n} = (x_s - x_c)n_x + (y_s - y_c) \cdot n_y$$

   if $\xi^3 > 0$ then exit the element.
3. Compute the normal force

$$N = \varepsilon_N \xi^3$$

4. Compute the tangential displacement

$$\Delta \xi^1 = (\boldsymbol{u}_s - \boldsymbol{u}) \cdot \boldsymbol{\tau}$$

5. Compute the trial force

$$T^{trial} = -\varepsilon_T \Delta \xi^1$$

6. Perform the return-mapping scheme (equation (9.40)) in order to compute the real force
7. Compute within the return-mapping scheme the corresponding residual $[R_T]$ for frictional force
8. Compute within the return-mapping scheme the corresponding tangent matrix $[K_T]$ for the real frictional force
9. Compute the normal residual $[R_N]$
10. Compute the normal tangent matrix $[K_N]$

11. Compute the full residual

$$[R] = [R_N] + [R_T]$$

12. Compute the full tangent matrix

$$[K] = [K_N] + [K_T].$$

Remark 9.4.2 *The tangent matrix $[K]$ is symmetric for the sticking part and unsymmetric for the sliding part.*

Remark 9.4.3 *The initial condition for the tangential force (position of the "initial sticking point" is automatically fulfilled.*

Chapter 21 in Part II contains all necessary information for the implementation, verification and further examples for the frictional NTN contact element.

9.5 Frictional Node-To-Segment (NTS) Contact Element

The frictional Node-To-Segment (NTS) contact element possesses the same kinematics studied for non-frictional NTS element in Section 7.4. We recall here only one restriction for the NTS approach:

- For the discretization of the slave part, only finite elements with linear approximations are used.

Due to the possibility of large sliding, the major difference from the NTN approach is the correct computation of the tangential measure, which is discussed in detail in Section 9.2. Thus, for correct computation, the coordinate of initial sticking point ξ_0^1 should be defined at the beginning of computation without loading and stored for further computation. If the reversible load is necessary for computation then the update algorithm should be implemented for the case of sliding, as discussed in Section 9.2.2. Since the tangent vector ρ_1 is computed from the geometry of NTS element and, in general, has arbitrary length, the return-mapping scheme is taken in the full and not reduced form as in equation (9.11), also the weak form is taken in the full form in equation (9.27) and is adjusted for NTS, as discussed within the numerical integration for the NTS approach in Section 7.1.2.

9.5.1 Linearization and Discretization for the NTS Frictional Contact Element

The linearization steps of the frictional part shown in equation (9.31), together with the numerical integration for the NTS approach in Section 7.1.2, lead to the following:

$$\begin{aligned} D_v(\delta W_T^c) &= D_v(T^1)(\rho_{\xi^1} \cdot \rho_{\xi^1})\delta\xi^1 \\ &+ T^1 D_v(\rho_{\xi^1} \cdot \rho_{\xi^1})\delta\xi^1 \\ &+ T^1(\rho_{\xi^1} \cdot \rho_{\xi^1})D_v(\delta\xi^1). \end{aligned} \quad (9.56)$$

The full transformation includes the following steps performed, however, in 3D and then reduced to 2D cylindrical geometry:

- Linearization of the tangential force is performed in the form of covariant derivative $D_v(T^1)$ with regard to the return-mapping algorithm separately for sticking and sliding cases;
- Linearization of further geometrical parameters (such as $D_v(\delta\xi^1)$) is performed in the form of the covariant derivative, see the discussion in Section 6.5 and equations (6.38a, 6.38b and 6.38c).

The full linearization lays outside of scope of this text book and advanced readers are recommended to read the book by Konyukhov and Schweizerhof (2012). For further discretization, the approximation matrix $[A]$ for the NTS approach in equation (7.17) and the derivative of the approximation matrix $[A_{\xi^1}]$ in equation (7.18) are involved. We only represent here the final result for sticking and sliding cases separately.

Part of the tangent matrix for sticking:

$$[K_T^{stick}] = -\frac{\varepsilon_T}{\rho_{\xi^1} \cdot \rho_{\xi^1}}[A]^T \rho_{\xi^1} \otimes \rho_{\xi^1}[A]$$
$$- \frac{T_1^{real}}{(\rho_{\xi^1} \cdot \rho_{\xi^1})^2}\left[[A_{\xi^1}]^T \rho_{\xi^1} \otimes \rho_{\xi^1}[A] + [A]^T \rho_{\xi^1} \otimes \rho_{\xi^1}[A_{\xi^1}]\right] \quad (9.57)$$

Part of the tangent matrix for sliding:

$$[K_T^{slide}] = -\frac{\varepsilon_N \, \mu \, sign(T_1^{tr})}{\sqrt{\rho_{\xi^1} \cdot \rho_{\xi^1}}} \cdot [A]^T (\rho_{\xi^1} \otimes \mathbf{n}[A])$$
$$+ \frac{\mu|N|sign(T_1^{tr})}{(\rho_{\xi^1} \cdot \rho_{\xi^1})^{\frac{3}{2}}}\left[[A_{\xi^1}]^T \rho_{\xi^1} \otimes \rho_{\xi^1}[A] + [A]^T \rho_{\xi^1} \otimes \rho_{\xi^1}[A_{\xi^1}]\right] \quad (9.58)$$

9.5.2 Algorithm for a Local Level NTS Frictional Contact Element

We list here schematically the necessary steps for the implementation of the NTS frictional element, Chapter 22 in Part II contains all necessary informations for the implementation, verification and further typical examples of the NTS approach with the Coulomb friction law based on the Penalty method.

9.6 NTS Frictional Contact Element

1. Initialization of "the initial sticking point" with zero external load-compute $\xi_c^{1(0)}$ via the CPP procedure
2. Update coordinates
3. Compute the current projection $\xi^{1(n+1)}$ (CPP procedure)

4. Local search: if $|\xi^{1(n+1)}| < 1$ then exit
5. Penetration check: compute $\xi^3 = (\mathbf{r}_s - \boldsymbol{\rho}) \cdot \mathbf{n}$ if $\xi^3 > 0$ then exit
6. Compute normal force N and normal tangent matrix $[K_N]$
7. Perform the return-mapping scheme (equation (9.40)) in order to compute the real tangential force T^1_{real}
8. Compute within the return-mapping scheme the corresponding residual $[R_T]$ for frictional force
9. Compute within the return-mapping scheme the corresponding tangent matrices $[K_T]$ for the real frictional force
10. Update "the initial sticking point" within the return-mapping scheme, see Section 9.2.2, if sliding is detected (only necessary for reversible load)
11. Compute the normal residual $[R_N]$
12. Compute the normal tangent matrix $[K_N]$
13. Compute full residual
$$[R] = [R_N] + [R_T]$$
14. Compute the full tangent matrix
$$[K] = [K_N] + [K_T].$$

Remark 9.6.1 *The initial condition for the tangential force (position of the "initial sticking point" is defined by performing the global solution without the initial loading. In this case, the result of the CPP procedure gives us the position of the initial sticking point in which the initial tangent force is assumed to be zero. This a major difference from the simplest NTN approach, compared with Section 9.4.4.*

Chapter 22 in Part II contains all necessary information for the implementation, verification and further examples for the frictional NTS contact element.

Remark 9.6.2 *For high-order approximations of the contact segments (classical high-order finite element, spline finite elements, isogeometric finite elements) it is necessary to implement all parts of the tangent matrix (including the curvature part) in order to keep the quadratic rate of convergence for the Newton method. Also, more careful treatment is necessary for the update of history variables scheme discussed in Section 9.2.2.*

As we know, the NTS approach is still applicable in the case of the high-order approximation for the master segment only if the slave segment is rigid, see Remark 7.4.4. Namely, this class of examples are chosen in order to show peculiarities within the implementation of the high-order approximation for the frictional NTS approach and this is illustrated in Chapter 23 in Part II.

Part II

Programming and Verification Tasks

10

Introduction to Programming and Verification Tasks

Part II of the book concentrates on the direct application of the theoretical part described before. The reader shall gain insight into the implementation of various finite elements as well as their application to examples. The focus concentrates on elements regarding nonlinearities mostly due to contact, however, a few necessary structural nonlinear elements are also discussed.

Since the exercise part of this book emerged from lectures on this field to postgraduate classes, the style and content reflects the attributes and abilities gained within these courses. The structure of these classes can be easily seen by the amount of lessons each representing a class of three hours. The topics discussed first of all concentrate on structural elements as truss and plane stress elements (`elmt1.f` - `elmt2.f`) that are used further within the contact mechanics examples. Readers with a background of implementation in nonlinear structural finite elements may skip this part. Their focus may be more on the following contact elements with different contact formulations such as the Penalty, the Lagrange multiplier and Nitsche methods (`elmt100.f` - `elmt102.f`) with Node-To-Node (NTN) contact discretization in 1D. Further focus is on the contact discretization in 2D with the Node-To-Segment (NTS), Segment-To-Analytical-Segment (STAS) and Segment-To-Segment (STS) with linear and higher order isoparametric approximation and an extension to 3D problems (`elmt103.f` - `elmt107.f`). At this stage the reader should be capable of performing any combination of discretization with any contact enforcement, even for 3D problems. The next block then consists of contact elements for 1D and 2D frictional problems (`elmt108.f` - `elmt110.f`). The last section demonstrates the applicability of the derived contact elements to transient contact problems.

As discussed within Chapter 5 of Part I, the contact problem leads to nonlinear equilibrium equations. One of the most popular and simplest methods to solve nonlinear

---

Introduction to Computational Contact Mechanics: A Geometrical Approach, First Edition.
Alexander Konyukhov and Ridvan Izi.
© 2015 John Wiley & Sons, Ltd. Published 2015 by John Wiley & Sons, Ltd.
Companion Website: www.wiley.com/go/Konyukhov

equations is the Newton iterative method described in detail within Section 5.4 in Part I, in which the applied external load is raised in increments of in. Each implementation of a contact element, starting with various discretization approaches described in Chapter 7 of Part I, requires evaluation of the tangent matrix $[s]$ and residual vector $[p]$. Implementation of them gives a necessary set defining a certain contact finite element (NTN, NTS, STAS, STS, etc.). In order to achieve the same level as within the postgraduate classes it is expected that the reader already has some familiarity with computer implementation of the finite element method in the linear context. In this context, it is essential to understand two crucial routines. Firstly, the master routine that controls the overall organization of the program and secondly, the element subroutine elmtxx.f. This general setup of a finite element program is presented in Figure 10.1 showing a generalized flowchart of such a finite element code. The only part of this flowchart that the exercise part of this book takes care of is the element level where the tangent matrix and residual vector for each element is computed and provided to the main program. The master routines deal with reading the input file regarding the setup of the problem, allocating the required memory, reading the nodal coordinates, materials, boundary conditions and loading. Furthermore, it assembles all element matrices and vectors and solves the algebraic system. As can be seen, the only unknowns within this flowchart are the displacement values at each node $[ul]$. For all problems, the stop criteria for the iterative solution procedure become the incremental energy consisting of incremental displacements of the iteration and the corresponding tangent matrix. Once converging the iteration gets closer to its final solution with incremental energy declining below a given tolerance of tol, finally computation stops for this load step.

Particular attention is given to the discussion of the fully structural case – a special case of the following normal forces, such as a single force, or distributed forces (inflating pressure). In this case the point of view can be regarded as a test to check the correctness of implementation for the rotational part, however, from another point of view it can be regarded as an *inverted contact algorithm* for the implementation of the following forces. The following forces as an *inverted contact algorithm* can be implemented based on various contact approaches NTS, STAS and STS. This implementation, first of all, is regarded as a verification test for the rotational part (and in addition curvature part in case of high order finite elements). However, each contact approach (NTS, STAS or STS) allows a certain mechanical interpretation as following normal single or distributed forces.

Since the focus is on the implementation of each element, every element subroutine is presented in form of a flowchart enabling the implementation in any numerical computation environment as MATLAB, Maple or Mathematica as well as in any finite element analysis program based on different programming languages such as Fortran or C. The environment used by the authors is based on a finite element analysis program called FEAP, originated by Professor Taylor, which is designed for research and educational use. Source code of the full program is available for compilation using Windows, LINUX or UNIX operating systems. A small version of the system,

Introduction to Programming and Verification Tasks

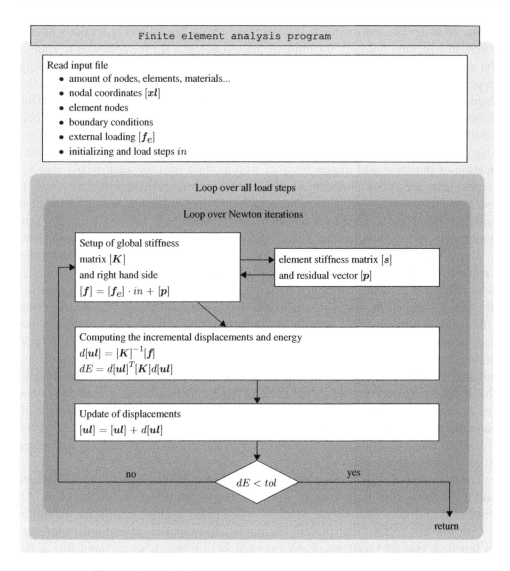

Figure 10.1 Global setup of a finite element analysis program

called FEAPpv, is available free of charge. See *www.ce.berkeley.edu/feap/feappv* for additional information. The version used by the authors is FEAP-MeKa and can be regarded as a further development of an earlier version of FEAP, which enables the material input $[d]$ and setting up of the tangent matrix $[s]$ and the residual vector $[p]$ within the same element. Furthermore, a post-processing with LS-PrePost (*www.lstc.com/lspp*) is enabled within this version. FEAP-MeKa, together with user and programmer instructions for the LINUX operating system, templates for programming the presented elements, and the encountered examples can be obtained from the publishers via accessing the site www.wiley.com/go/Konyukhov.

Due to using this environment, all elements within the following sections have a common structure, not only for implementation but also for the followed testing of the element within finite element analysis. First, all elements are described briefly in general and presented in a figure demonstrating the setup of all nodes where the displacements are denoted by u and v for the x- and y- directions, respectively. Due to dealing with nonlinear elements in general, all elements require an update of their position ($x = X + u$) and use, unless otherwise specified, linear shape functions of the Lagrange class in isoparametric sense. For contact elements, additionally, the identification of the slave and master body as well as their position vectors ρ_M and r_S, respectively, are given. Furthermore, the defined normal and tangential direction n and τ and their corresponding coordinates ξ^3 and ξ^1 are presented. Secondly, the general description is followed by a flowchart for the element to be programmed where major steps are summarized in subroutines and loops over, for example, integration points, that are given in shaded boxes. These subroutines are then further explained and described with reference to the theoretical part. Here, all shown calculations are matrix or vector multiplications, the approximation matrix and its derivatives are stored within the matrices $[a]$ and $[a_{,\xi^i}]$, respectively. All global FEAP-arrays specific for each element as the material parameters $[d]$, the initial nodal coordinates $[xl]$ and the nodal displacements $[ul]$ are given. The material parameters $[d]$ are defined during the application of the element for a finite element analysis by the user and are different for each finite element. The reader is highly advised to work closely with the FEAP-MeKa code provided (written in FORTRAN) where a programmed version for all discussed elements is provided within the supporting web page for download. Nevertheless, the presented flowchart structure and the following explanations for the subroutines should enable readers using other various programming techniques to gain the intended results.

The last point of each exercise section consists of examples making the advantages and disadvantages of each contact formulation and contact discretization clear. All discussed examples demonstrate static or quasi-static loading, except the examples within the last section that present the applicability of the programmed elements to transient contact problems. Each example is provided with its input-file in the earlier-mentioned downloadable folder in which the structure is comparable to the exercise part of this book. Besides, the input-files some example folders contain additional material necessary to gain the presented results of the computation. Hereby, the expression post-processing denotes the use of the graphical visualization of LS-PrePost with the generated file intau, plot intends to achieve a diagram using gnuplot – some cases require an AWK- or Fortran-script to extract the necessary data for this.

Furthermore, some examples require additional structural elements, such as the geometrically and physically linear spring, truss and plane elements, higher order structural nonlinear plane and volume elements. All these elements are provided within the given finite element code, but are not further specified.

11

Lesson 1
Nonlinear Structural Truss – `elmt1.f`

The reader is prompted to start with the structural element possessing the most simplest geometry describing truss structures in geometrically nonlinear formulations and allowing both linear elastic (St. Venant) and nonlinear elastic (Neo-Hooke) material law.

The nonlinear structural truss element (`elmt1.f`) with linear shape functions consists of two nodes in the 2D plane as can be seen in Figure 11.1. The formulation used is according to the theory given in Bonet and Wood (1999) and includes in addition a uniformly distributed normal force n. Further, the material parameter E describes the elasticity modulus and A the area of cross section. Since the distributed normal force in general requires a numerical integration, the Gauss quadrature is considered.

In order to use this truss element in 2D, an additional transformation from 2D properties to 1D and back whilst not influencing the physical property of the element, is performed.

The truss finite element can be used together with contact elements possessing the simplest contact kinematics, such as Node-to-Node (NTN). A reader experienced in nonlinear structural finite element analysis, who is interested in contact finite elements, can skip these first two chapters in Part II.

11.1 Implementation

Setup of tangent matrix and residual (`isw = 3`)

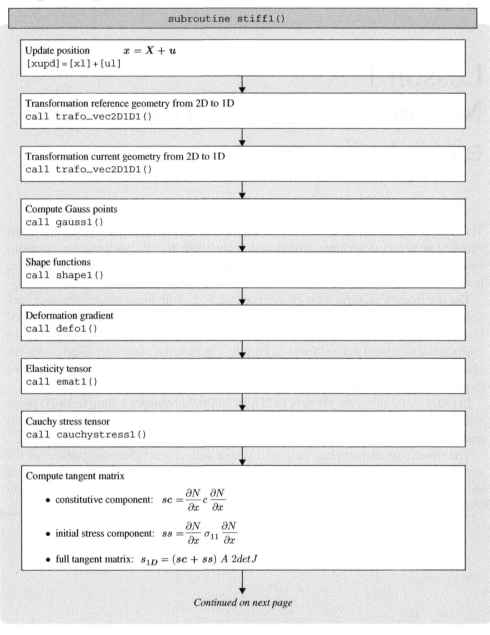

Lesson 1 Nonlinear Structural Truss – elmt1.f

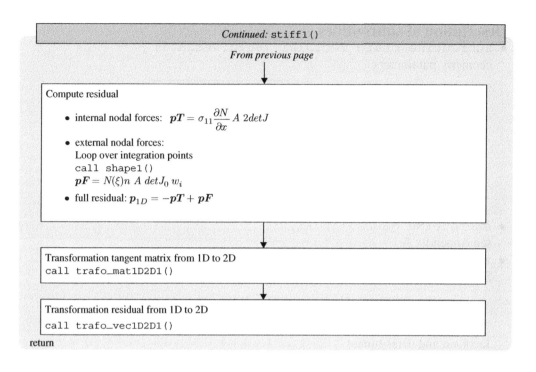

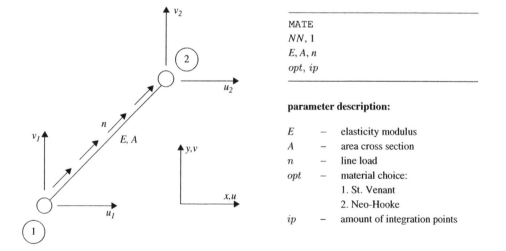

Figure 11.1 Geometry and parameters of the nonlinear truss finite element

Description of subroutines
- subroutine trafo_vec2D1D1(vec1d,vec2d,l)
  geometry parameters
  $$l = \sqrt{(x_2 - x_1)^2 + (y_2 - y_1)^2}$$
  $$c_x = \frac{x_2 - x_1}{l}, \quad s_x = \frac{y_2 - y_1}{l}$$
  $$[\text{trafo}] = \begin{bmatrix} c_x & s_x & 0 & 0 \\ 0 & 0 & c_x & s_x \end{bmatrix}$$

  $[\text{vec1d}] = [\text{trafo}][\text{vec2d}]$

- subroutine gauss1(ip,xi,wi)
  see Appendix A
- subroutine shape1(shape,xi,l,detJ)
  Shape functions N_i and local derivatives $N_{i,\xi}$
  $$N_1 = \tfrac{1}{2}(1-\xi) \quad N_2 = \tfrac{1}{2}(1+\xi)$$
  $$N_{1,\xi} = -\tfrac{1}{2} \quad N_{2,\xi} = \tfrac{1}{2}$$
  Jacobian and determinant
  $$J = \det J = \frac{l}{2}$$
  Global derivatives of shape functions N_i
  $$N_{1,x} = J^{-1} N_{1,\xi} \quad N_{2,x} = J^{-1} N_{2,\xi}$$

- subroutine defo1(F_{11},detF,xupd,xl)
  Deformation gradient and determinant
  $$F_{11} = \frac{\partial x}{\partial X} = \frac{xupd(2) - xupd(1)}{xl(2) - xl(1)}$$
  $$\det \boldsymbol{F} = F_{11}$$

- subroutine emat1(c,d,detF)
  Elasticity tensor
  call dmat1(λ,μ,d)

  St. Venant:
  if (opt.eq.1) then
  $c = \lambda + 2\mu$

  Neo-Hooke:
  elseif (opt.eq.2) then
  $$c = \frac{\lambda}{\det \boldsymbol{F}} + \frac{2(\mu - \lambda \ln(\det \boldsymbol{F}))}{\det \boldsymbol{F}}$$

- subroutine dmat1(λ,μ,d)

  $$\lambda = 0, \quad \mu = \frac{E}{2}$$

Lesson 1 Nonlinear Structural Truss – elmt1.f

- subroutine cauchystress1(sig,F11,detF,d)
  Cauchy stress tensor
  call dmat1(λ,μ,d)

St. Venant:
if (opt.eq.1) then
$$\sigma_{11} = F_{11}\left[\frac{\lambda}{2}(F_{11}F_{11}-1) + \mu(F_{11}F_{11}-1)\right]$$

Neo-Hooke:
elseif (opt.eq.2) then
$$\sigma_{11} = \frac{\mu}{\det \boldsymbol{F}}(F_{11}F_{11}-1) + \frac{\lambda}{\det \boldsymbol{F}}\ln(\det \boldsymbol{F})$$

- subroutine trafo_mat1D2D1(mat1d,mat2d,xupd)
  geometry parameters
$$l = \sqrt{(x_2-x_1)^2 + (y_2-y_1)^2}$$
$$c_x = \frac{x_2-x_1}{l}, \quad s_x = \frac{y_2-y_1}{l}$$
$$[\text{trafo}] = \begin{bmatrix} c_x & s_x & 0 & 0 \\ 0 & 0 & c_x & s_x \end{bmatrix}$$

$[\text{mat2d}] = [\text{trafo}]^T[\text{mat1d}][\text{trafo}]$

- subroutine trafo_vec1D2D1(vec1d,vec2d,xupd)
  geometry parameters
$$l = \sqrt{(x_2-x_1)^2 + (y_2-y_1)^2}$$
$$c_x = \frac{x_2-x_1}{l}, \quad s_x = \frac{y_2-y_1}{l}$$
$$[\text{trafo}] = \begin{bmatrix} c_x & s_x & 0 & 0 \\ 0 & 0 & c_x & s_x \end{bmatrix}$$

$[\text{vec2d}] = [\text{trafo}]^T[\text{vec1d}]$

Global *FEAP*-arrays

$[\text{d}] = \begin{bmatrix} E & A & n & opt & ip \end{bmatrix}$ (material parameters)

$[\text{xl}] = \begin{bmatrix} x_1 & x_2 \\ y_1 & y_2 \end{bmatrix}$ (nodal coordinates)

$[\text{ul}] = \begin{bmatrix} u_1 & u_2 \\ v_1 & v_2 \end{bmatrix}$ (nodal displacements)

Hints for implementation

1. Entries of $[s]$ have to be symmetric as $[s]^T = [s]$ is valid.
2. Residual set as $[s][ul]$ leads also to a converging computation, but with more iterations.

11.2 Examples

The examples presented in the following are intended to show the difference between a physically and geometrically linear and nonlinear truss element. For this purpose the provided finite element code includes a physically and geometrically linear 2D truss element (elmt3.f).

11.2.1 Constitutive Laws of Material

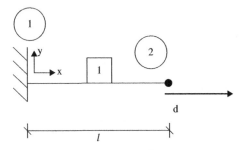

The first problem shows a truss element under tensile or compression load. A horizontal displacement d is applied in increments (in). The goal is to study the physical nonlinearity. Compare the element when a St. Venant material choice or Neo-Hooke material choice is made and plot the force/displacement diagram for compression and tension.

Parameters: $E = 10^4, A = 0.1, l = 1, d = -1/6, in = 50/120$

Results:

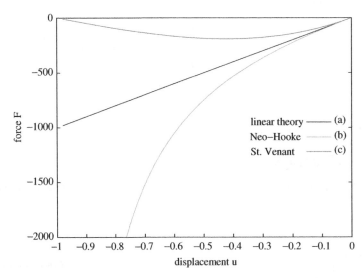

Figure 11.2 Compression of truss element. Comparison: (a) physically and geometrically linear theory, (b) Neo-Hooke and (c) St. Venant and geometrically nonlinear theory

In this diagram Fig. 11.2 force vs displacement during compression of the bar is shown for various material laws: geometrically linear, Neo-Hooke and St. Venant types of material.

Lesson 1 Nonlinear Structural Truss – elmt1.f

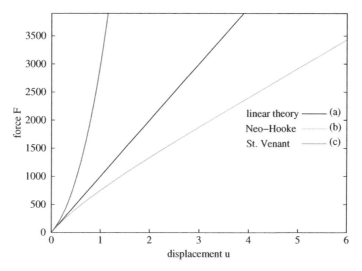

Figure 11.3 Tension of truss element. Comparison: (a) physically and geometrically linear theory, (b) Neo-Hooke and (c) St. Venant and geometrically nonlinear theory

In this diagram Fig.11.3 force vs displacement during tension of the bar is shown for various material laws: geometrically linear, Neo-Hooke and St. Venant types of material.

11.2.2 Large Rotation

Two hinged trusses modelling a simple mechanism are subjected to large rotation during loading. The goal is to study the geometrical nonlinearity. With a linear approximation it is not possible to describe large rotations. On the contrary, a nonlinear finite element approximation can describe large rotations correctly.

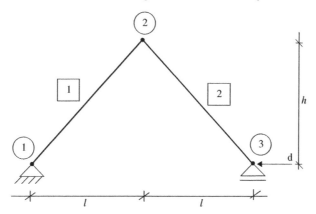

In this figure, a setup is shown for a horizontal displacement d applied in increments (in). The problem is solved using a linear (elmt3.f) and a nonlinear (elmt1.f) element. What difference can be seen within the postprocessing?

Parameters: $E = 5 \cdot 10^6$, $A = 1$, $l = 10$, $h = 1$, $d = 20$, $in = 400$

11.2.3 Snap-Through Buckling

The third example shows the famous snap-through buckling example, which is only correctly solved with geometrically nonlinear finite elements.

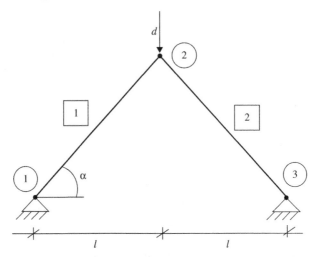

On the setup of two truss elements, a vertical displacement d is applied in increments (in). The problem is solved using the nonlinear element (elmt1.f). Plot the force/displacement for different angles α ($\alpha = 15, 20, 25, 30, 35, 40, 45$).

Parameters: $E = 10^6$, $A = 1$, $l = 10$, $d = -15$, $in = 150$

Results of computation are given in Fig. 11.4:

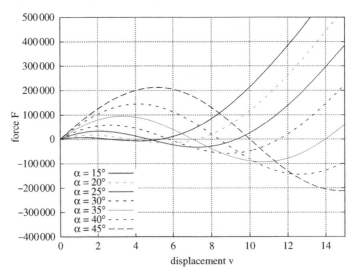

Figure 11.4 Force displacement diagram for snap-through buckling (Neo-Hooke)

Diagram in Fig. 11.4 is illustrating force vs displacement behavior for various initial angle for snap-through buckling, see Section 11.2.3.

12

Lesson 2
Nonlinear Structural Plane – `elmt2.f`

The nonlinear structural plane element (`elmt2.f`) with linear shape functions consists of four nodes as can be seen in Figure 12.1. The derivation here is again according to the theory given in Bonet and Wood (1999) and is even including a consistent mass matrix which is further used for the transient problems in Chapter 24. Thus, the density ρ is a parameter to choose, as well as the structural parameters E, ν and t. Furthermore, a volumetric loading in x- and y-direction is enabled with q_x and q_y.

The nonlinear structural plane is the most used finite element within the current exercise part for the application to 2D contact problems. The element is implemented for a Neo-Hooke material law with a geometrically nonlinear formulation.

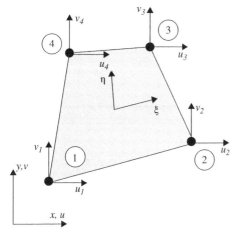

MATE
$NN, 2$
E, ν, t, ip, q_x, q_y
ρ

parameter description:

E	–	elasticity modulus
ν	–	Poisson's ratio
t	–	thickness
ip	–	amount of integration points ($ip=2$)
q_x	–	volume loads in x-direction
q_y	–	volume loads in y-direction
ρ	–	density

Figure 12.1 Geometry and parameters of the nonlinear plane finite element

Introduction to Computational Contact Mechanics: A Geometrical Approach, First Edition.
Alexander Konyukhov and Ridvan Izi.
© 2015 John Wiley & Sons, Ltd. Published 2015 by John Wiley & Sons, Ltd.
Companion Website: www.wiley.com/go/Konyukhov

12.1 Implementation

Setup of tangent matrix and residual (`isw = 3`)

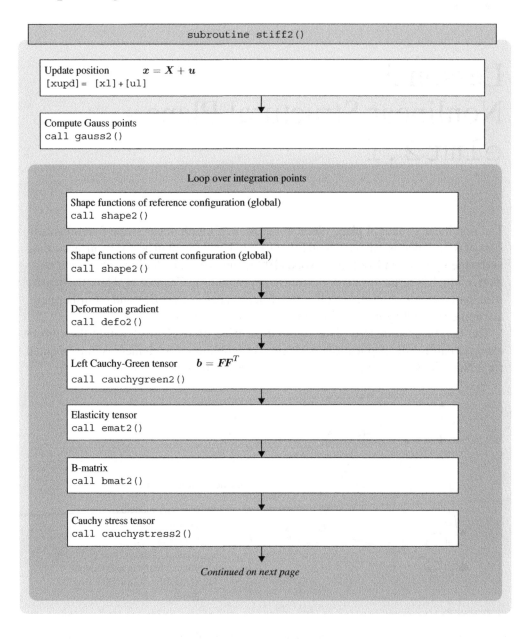

Lesson 2 Nonlinear Structural Plane – `elmt2.f`

> *Continued:* `stiff2()`
> *From previous page*
>
> **Compute tangent matrix**
> - constitutive component: $sc = B^T DB$
> - initial stress component: $ss = \begin{bmatrix} [K_{11}] & \cdots & [K_{14}] \\ \vdots & & \vdots \\ [K_{41}] & \cdots & [K_{44}] \end{bmatrix}$
>
>   with $K_{ab} = \dfrac{\partial N_a}{\partial x_k} \sigma_{kl} \dfrac{\partial N_b}{\partial x_l} \begin{bmatrix} 1 & 0 \\ 0 & 1 \end{bmatrix}$
> - full tangent matrix: $s = (sc + ss) \cdot det J \cdot wi_i \cdot wi_j \cdot t$
>
> **Compute residual**
> - internal nodal forces: $pT = B^T \sigma_V \cdot det J \cdot wi_i \cdot wi_j \cdot t$
> - external nodal forces: $pF = \begin{bmatrix} F_1 \\ \vdots \\ F_4 \end{bmatrix} \cdot det J_0 \cdot wi_i \cdot wi_j \cdot t$
>
>   with $F_a = N_a \begin{bmatrix} q_x \\ q_y \end{bmatrix}$
> - full residual: $p = -pT + pF$
>
> return

Setup of mass matrix (`isw = 5`)

> `subroutine mass2()`
>
> **Update position** $\quad x = X + u$
> `[xupd] = [xl] + [ul]`
>
> **Compute Gauss points**
> `call gauss2()`
>
> **Loop over integration points**
>
> > **Shape functions of current configuration (global)**
> > `call shape2()`
> >
> > **Consistent shape functions**
> > $$N = \begin{bmatrix} N_1 & 0 & \cdots & N_4 & 0 \\ 0 & N_1 & \cdots & 0 & N_4 \end{bmatrix}$$
> >
> > **Compute mass matrix**
> > $s = \rho N^T N \cdot det J \cdot wi_i \cdot wi_j \cdot t$
>
> return

Description of subroutines

- `subroutine gauss2 (ip,xi,wi)`
  see Appendix A
- `subroutine shape2 (shape,shpxy,detj,xi,eta,xl)`
  Shape functions N_i and local derivatives $N_{i,\xi}$, $N_{i,\eta}$

$$[\text{shape}] = \begin{bmatrix} N_1 & N_2 & N_3 & N_4 \\ N_{1,\xi} & N_{2,\xi} & N_{3,\xi} & N_{4,\xi} \\ N_{1,\eta} & N_{2,\eta} & N_{3,\eta} & N_{4,\eta} \end{bmatrix}$$

with $N_i = \frac{1}{4}(1 + \xi\xi_i) \cdot (1 + \eta\eta_i)$ and

i	1	2	3	4
ξ_i	-1	1	1	-1
η_i	-1	-1	1	1

Jacobian, determinant and inverse

$$[\text{jm}] = \begin{bmatrix} \frac{\partial x}{\partial \xi} & \frac{\partial y}{\partial \xi} \\ \frac{\partial x}{\partial \eta} & \frac{\partial y}{\partial \eta} \end{bmatrix} = \sum_{i=1}^{4} \begin{bmatrix} N_{i,\xi}\, x_i & N_{i,\xi}\, y_i \\ N_{i,\eta}\, x_i & N_{i,\eta}\, y_i \end{bmatrix}$$

$$\text{detj} = jm_{11} \cdot jm_{22} - jm_{12} \cdot jm_{21}$$

$$[\text{jinv}] = \frac{1}{\text{detj}} \begin{bmatrix} jm_{22} & -jm_{12} \\ -jm_{21} & jm_{11} \end{bmatrix}$$

Global derivatives of shape functions N_i

$$\begin{bmatrix} N_{i,x} \\ N_{i,y} \end{bmatrix} = [\text{jinv}] \begin{bmatrix} N_{i,\xi} \\ N_{i,\eta} \end{bmatrix}$$

- `subroutine defo2 (F,detF,shpxy,xupd)`
  Deformation gradient and determinant

$$\boldsymbol{F} = \begin{bmatrix} N_{i,x} \\ N_{i,y} \end{bmatrix} [\text{xupd}]^T$$

$$\text{det}\boldsymbol{F} = F_{11} \cdot F_{22} - F_{12} \cdot F_{21}$$

- `subroutine cauchygreen2 (b,F)`
  Left Cauchy–Green tensor

$$\boldsymbol{b} = \boldsymbol{F}\,\boldsymbol{F}^T$$

Lesson 2 Nonlinear Structural Plane – elmt2.f

- `subroutine emat2 (dmat,detF,d)`
  Elasticity tensor Neo-Hooke (Voigt notation)

$$[\text{dmat}] = \begin{bmatrix} \overline{\lambda} + 2\cdot\overline{\mu} & \overline{\lambda} & 0 \\ \overline{\lambda} & \overline{\lambda} + 2\cdot\overline{\mu} & 0 \\ 0 & 0 & \overline{\mu} \end{bmatrix}$$

with effective parameters

$$\lambda = \frac{\nu \cdot E}{(1+\nu)\cdot(1-2\cdot\nu)} \quad \text{and} \quad \mu = \frac{E}{2\cdot(1+\nu)}$$

$$\overline{\lambda} = \frac{\lambda}{\det F} \quad \text{and} \quad \overline{\mu} = \frac{\mu - \lambda\cdot\ln(\det F)}{\det F}$$

- `subroutine bmat2 (bmat,shpxy)`
  Global derivatives of shape functions in **B**-matrix

$$\boldsymbol{B} = \begin{bmatrix} [\text{b}]_1 & [\text{b}]_2 & [\text{b}]_3 & [\text{b}]_4 \end{bmatrix}$$

with

$$[\text{b}]_i = \begin{bmatrix} N_{i,x} & 0 \\ 0 & N_{i,y} \\ N_{i,y} & N_{i,x} \end{bmatrix}$$

- `cauchystress2 (sig,sigV,b,detF,d)`
  Cauchy stress tensor and vector (Voigt notation) for Neo-Hooke material choice

Stress tensor

$$\boldsymbol{\sigma} = \frac{\mu}{\det F}\cdot(\boldsymbol{b}-\boldsymbol{1}) + \frac{\lambda}{\det F}\cdot\ln(\det F)\cdot\boldsymbol{1}$$

Stress vector (Voigt notation)

$$\boldsymbol{\sigma}_V = \begin{bmatrix} \sigma_{11} \\ \sigma_{22} \\ \sigma_{12} \end{bmatrix}$$

Global *FEAP*-arrays

$$[\text{d}] = \begin{bmatrix} E & \nu & t & ip & q_x & q_y & \rho \end{bmatrix} \quad \text{(material parameters)}$$

$$[\text{xl}] = \begin{bmatrix} x_1 & x_2 & x_3 & x_4 \\ y_1 & y_2 & y_3 & y_4 \end{bmatrix} \quad \text{(nodal coordinates)}$$

$$[\text{ul}] = \begin{bmatrix} u_1 & v_2 & u_3 & v_4 \\ v_1 & v_2 & v_3 & v_4 \end{bmatrix} \quad \text{(nodal displacements)}$$

Hints for implementation

1. Entries of $[s]$ have to be symmetric as $[s]^T = [s]$ is valid.
2. Residual set as $[s][ul]$ leads also to a converging computation, but with more iterations.

12.2 Examples

As in case of the nonlinear truss of Chapter 11 the following examples are intending to present the difference between a geometrically and physically linear and nonlinear plane element. For this purpose the finite element code provides a fully linear plane element (elmt4.f).

12.2.1 Constitutive Law of Material

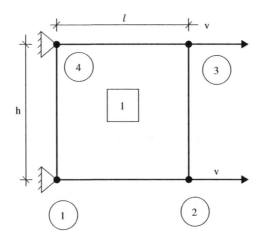

The first problem shows a plane element under tensile or compression load. A horizontal displacement v is applied in increments (in). Plot the force/displacement diagram for compression and tension.

Parameters: $E = 10^4, \nu = 0, h = 1, l = 1, t = 0.1, ip = 2, v = -1/6, in = 50/120$

Results of computation are given in Figs. 12.2 and 12.3:

Lesson 2 Nonlinear Structural Plane – elmt2.f

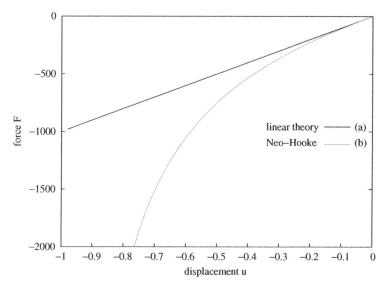

Figure 12.2 Compression of plane element. Comparison: (a) physically and geometrically linear theory, (b) Neo-Hooke and geometrically nonlinear theory

During compression of the element with non-linear law (Neo-Hooke) the force is increasing in magnitude highly-nonlinear, while for linear law the linearity is remaining, see Fig. 12.2.

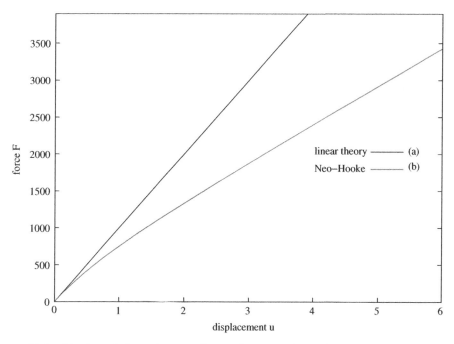

Figure 12.3 Tension of plane element. Comparison: (a) physically and geometrically linear theory, (b) Neo-Hooke and geometrically nonlinear theory

During tension of the element with non-linear law (Neo-Hooke) the force is increasing in magnitude slowly than linear, while for linear law the linearity is remaining, see Fig. 12.3.

12.2.2 Large Rotation

Two hinged plane elements modeling a simple mechanism are subjected to large rotation during loading. The goal is to study the difference between a linear and nonlinear finite element modeling with regards to the geometrical nonlinearity.

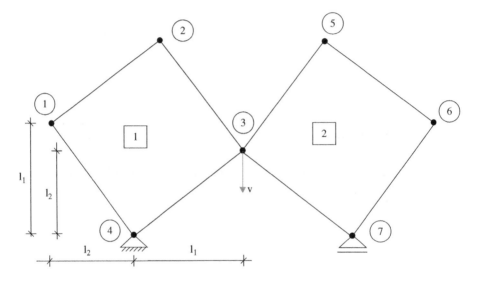

On the shown setup, a vertical displacement v is applied in increments (in). The problem is solved with linear (elmt4.f) and nonlinear (elmt2.f) plane elements. What difference can be seen with postprocessing?

Parameters: $E = 5 \cdot 10^6$, $\nu = 0$, $l_1 = 4$, $l_2 = 3$, $t = 1$, $ip = 2$, $v = 6$, $in = 120$

13

Lesson 3
Penalty Node-To-Node (NTN) – `elmt100.f`

This chapter starts the initial intention of the implementation of contact elements. As stated within the theoretical Part I, the contact elements considered can be based on different combinations of contact surface discretizations on 1D, 2D and 3D geometry and also on different enforcements of the non-penetration condition. Within this and the following two sections the focus is on having the simplest geometry and discretization in order to present different enforcement formulations. Thus, the following text will concentrate on the Node-To-Node (NTN) discretization successively using the Penalty, Lagrange multiplier and Nitsche formulations.

The NTN discretization is the simplest one with many restrictions, however, it still can be used in practice if its assumptions are fulfilled *a priori*. The theory and its assumptions are studied within Section 7.2 in Part I. In case of the Penalty formulation the NTN element consists of two finite element nodes where, besides the normal vector $\mathbf{n}$, the Penalty parameter in the normal direction ε_N has to be specified by the user as a material parameter. Although, `elmt100.f` is a very short element to program, its structure will be based on a common pattern where major programming steps are summarized within `subroutines`. This will enable a more easy approach regarding the differences between the proposed elements. Such a common point will be, for example, the position matrix $[a]$ expressing the distance between slave and master body as $\mathbf{r}_S - \boldsymbol{\rho}_M$. In this sense each contact element will have one specific position matrix, which differs with respect to size and entries for each contact element.

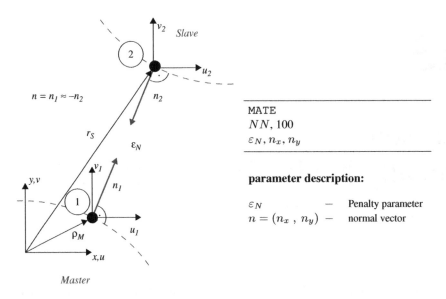

Figure 13.1 Geometry and parameters of the Penalty NTN contact element

13.1 Implementation

Setup of tangent matrix and residual (isw = 3)

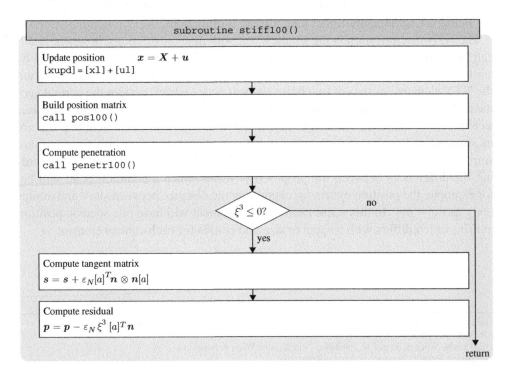

Lesson 3 Penalty Node-To-Node (NTN) – elmt100.f

Description of subroutines

- `subroutine pos100(a)`
  Position matrix for $(r_S - \rho_M) = [a]\,[xupd]$
  $$[a] = \begin{bmatrix} -1 & 0 & 1 & 0 \\ 0 & -1 & 0 & 1 \end{bmatrix}$$

- `subroutine penetr100(xupd,n,a,penetr)`
  Penetration
  $$\xi^3 = [xupd]^T\,[a]^T\,n$$

Global *FEAP*-arrays

$[\text{d}] = \begin{bmatrix} \varepsilon_N & n_x & n_y \end{bmatrix}$ (material parameters)

$[\text{xl}] = \begin{bmatrix} x_M & x_S \\ y_M & y_S \end{bmatrix}$ (nodal coordinates)

$[\text{ul}] = \begin{bmatrix} u_M & u_S \\ v_M & v_S \end{bmatrix}$ (nodal displacements)

Hints for implementation

1. Entries of $[s]$ have to be symmetric as $[s]^T = [s]$ is valid.
2. Residual set as $[s][ul]$ leads also to a converging computation, but with more iterations.

13.2 Examples

The following examples are intended for checking the correct implementation of the NTN discretization with a Penalty type of enforcement and giving an idea of possible fields of application.

13.2.1 Two Trusses

The first example as can be seen in the setup is dealing with two trusses where the left one is subjected to a force of F applied in increments (in).

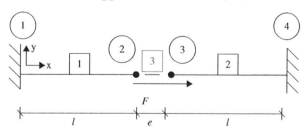

Parameters: $E = 10^3$, $A = 1$, $l = 5$, $e = 0.1$, $F = 50$, $in = 5$, $\varepsilon_N = 10^5$, $\boldsymbol{n} = \begin{pmatrix} 1 \\ 0 \end{pmatrix}$

13.2.2 Three Trusses

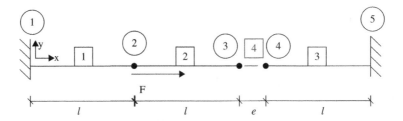

In the shown setup of three truss elements, a horizontal force F is applied in increments (in). Contact is checked between the trusses by a NTN approach. The goal is to use the implemented NTN Penalty contact element to specify the contact condition between trusses in the 1D case (nodes 3 and 4). Define the required contact elements (elmt100.f) to solve the contact problem. Study the influence of the penalty parameter ε_N on the value of penetration (Penetration is order of $\frac{1}{\varepsilon_N}$).

Parameters: $E = 10^3$, $A = 1$, $l = 5$, $e = 0.1$, $F = 50$, $in = 5$, $\varepsilon_N = 10^5$, $\boldsymbol{n} = \begin{pmatrix} 1 \\ 0 \end{pmatrix}$

Results of computation are given in Fig. 13.2:

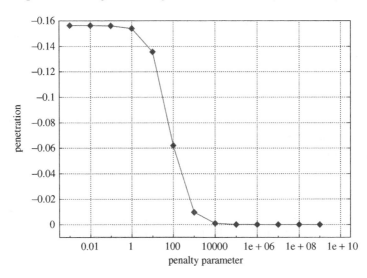

Figure 13.2 Influence of the penalty parameter ε_N on the value of penetration

The Fig. 13.2 is illustrating the behavior of the penetration: If the penalty parameter is small (close to zero), then the penetration is close to the applied displacement. If the penalty parameter is close to the elastic module E then penetration is close to zero (of order $1/E$). The larger value of penalty parameter is illustrating stabilization of the penetration.

13.2.3 Two Blocks

Contact between two blocks modeled by two structural plane elements is the simplest example in 2D enabling to verify kinematical properties such as the non-penetration condition of correctly implemented contact elements.

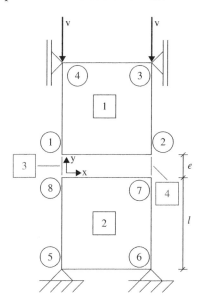

Here, a vertical displacement of v is applied in increments (in). Contact is checked between both planes by a NTN approach. Define the required contact elements (elmt100.f) to solve the contact problem. The goal is to study the influence of the penalty parameter ε_N on the value of penetration.

Parameters: $E = 10^5$, $\nu = 0.3$, $l = 1$, $v = -0.1$, $in = 10$, $e = 0.01$, $\varepsilon_N = 10^6$,
$$\boldsymbol{n} = \begin{pmatrix} 0 \\ 1 \end{pmatrix}$$

14

Lesson 4
Lagrange Multiplier Node-To-Node (NTN) – elmt101.f

The next formulation using the NTN discretization is the Lagrange multiplier method. Even though the method is originally based on Lagrange multipliers resulting in further unknowns besides the usual displacement values $[ul]$, the provided element setup waives this on behalf of the general setup of the exercise dealing only on an element level. In order to overcome this the Schur complement is applied to the whole element consisting of structural and contact parts. Thus, the programmed elements within Examples 14.2.1 and 14.2.2 are different and consist of four and five nodes, respectively. One can say that in this case the global tangent matrices are already assembled on element level within Figure 10.1. This means for the 2D case of Example 14.2.1 that the structural tangent matrix K and residual vector f consist of two truss elements resulting in 8×8 and 8×1 entries, respectively, where also the external loading $[f_e]$ is taken into account incrementally. For the application of the Lagrange multiplier method to the contact part in the sense of the Schur complement an assembly with taking the Dirichlet boundary conditions is necessary for the required inversions leading to K_{red}, f_{red} and $[a_{red}]$, see Remark 14.1.1. Nevertheless, as in the case of elmt100.f only the entries at the contacting area deal with the contact formulation, therefore, Figure 14.1 shows only two nodes for the contact. This type of implementation is just aimed at showing the Lagrange multipliers method on particular examples and by no means can be regarded as a general case of implementation.

Introduction to Computational Contact Mechanics: A Geometrical Approach, First Edition.
Alexander Konyukhov and Ridvan Izi.
© 2015 John Wiley & Sons, Ltd. Published 2015 by John Wiley & Sons, Ltd.
Companion Website: www.wiley.com/go/Konyukhov

14.1 Implementation

Setup of tangent matrix and residual (`isw = 3`)

---

subroutine stiff101()

Update position $x = X + u$
[xupd] = [xl] + [ul]

Compute structural part of stiffness matrix and residual vector

- Assembled tangent matrix with structural part ss (see Chapter 11 - s)

$$\text{e.g. } K = \begin{bmatrix} ss_1 & 0 \\ 0 & ss_2 \end{bmatrix}$$

- Assembled residual vector with structural part ps (see Chapter 11 - p)

$$\text{e.g. } f = \begin{bmatrix} ps_1 \\ ps_2 \end{bmatrix} + [f_e] \cdot in$$

Build position matrix
`call pos101()`

Compute penetration
`call penetr101()`

$\xi^3 \leq 0?$ no

yes

Compute full tangent matrix and residual (see Remark 14.1.1)

- full tangent matrix

$s = K$

- full residual

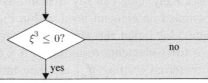

Full tangent matrix and residual
- $s = K$
- $p = f$

return

Lesson 4 Lagrange Multiplier Node-To-Node (NTN) – elmt101.f

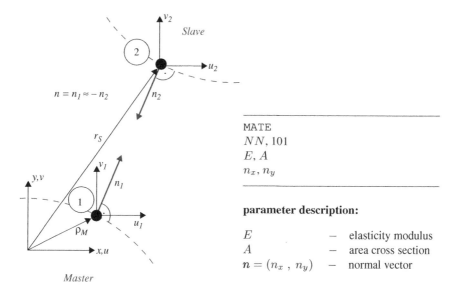

Figure 14.1 Geometry and parameters of the Lagrange multiplier NTN contact element

Description of subroutines

- subroutine pos101(a)
  Position matrix for $(r_S - \rho_M) = [a][xupd]$

$$[a] = \begin{bmatrix} -1 & 0 & 1 & 0 \\ 0 & -1 & 0 & 1 \end{bmatrix}$$

- subroutine penetr101(xupd,n,a,penetr)
  Penetration

$$\xi^3 = [xupd]^T [a]^T n$$

Remark 14.1.1 *Assembly of Lagrange formulation for Example 14.2.1 considering boundary conditions*

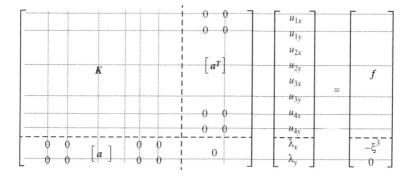

$$\Rightarrow \begin{bmatrix} \boldsymbol{K}_{red} & [a_{red}]^T \\ [a_{red}] & 0 \end{bmatrix} \begin{bmatrix} \boldsymbol{u}_{red} \\ \lambda_{red} \end{bmatrix} = \begin{bmatrix} \boldsymbol{f}_{red} \\ -\xi^3 \end{bmatrix} \quad (14.1)$$

Using the Schur complement on element level

$$\boldsymbol{K}\boldsymbol{u} = \boldsymbol{f} - [a_{red}]^T \left([a_{red}] \, \boldsymbol{K}_{red}^{-1} [a_{red}]^T\right)^{-1} \left([a_{red}] \, \boldsymbol{K}_{red}^{-1} \boldsymbol{f}_{red} + \xi^3\right)$$

Global *FEAP*-arrays

$$[\texttt{d}] = \begin{bmatrix} E & A & n_x & n_y \end{bmatrix} \qquad \text{(material parameters)}$$

$$[\texttt{xl}] = \begin{bmatrix} x_1 & \cdots & x_M & x_S & \cdots & x_{nen} \\ y_l & \cdots & y_M & y_S & \cdots & y_{nen} \end{bmatrix} \qquad \text{(nodal coordinates)}$$

$$[\texttt{ul}] = \begin{bmatrix} u_1 & \cdots & u_M & u_S & \cdots & u_{nen} \\ v_l & \cdots & v_M & v_S & \cdots & v_{nen} \end{bmatrix} \qquad \text{(nodal displacements)}$$

Hints for implementation

1. Entries of $[s]$ have to be symmetric as $[s]^T = [s]$ is valid.
2. Residual set as $[s][ul]$ leads also to a converging computation without contact enforcement.

14.2 Examples

The following examples deal with the NTN Lagrange element elmt101.f. Attention should be paid to the resulting penetration in comparison to the Penalty formulation, which should be zero – the non-penetration condition is fulfilled exactly for the Lagrange multiplier method, see the theoretical discussion in Section 1.3.1 in Part I.

14.2.1 Two Trusses

The first example, as can be seen in the setup, deals with two trusses where the left one is subjected to a force of F applied in increments (in). The setup is realized with one single element due to using the Schur complement for the Lagrange multiplier method. Thus, the element consists of four nodes and is denoted as elmt101a.f.

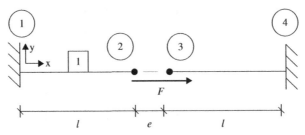

Parameters: $E = 10^3$, $A = 1$, $l = 5$, $e = 0.1$, $F = 50$, $in = 5$, $\boldsymbol{n} = \begin{pmatrix} 1 \\ 0 \end{pmatrix}$

14.2.2 Three Trusses

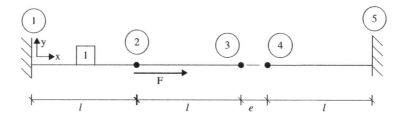

In the shown setup of three truss elements, a horizontal force F is applied in increments (in). Contact is checked between the trusses by a Lagrange multiplier NTN approach. The goal is to use the implemented NTN Lagrange multiplier contact element to specify the contact condition between the trusses in the 1D case (nodes 3 and 4). Define the required structural-contact element with all five nodes (elmt101b.f) to solve the contact problem. Study the value of penetration.

Parameters: $E = 10^3$, $A = 1$, $l = 5$, $e = 0.1$, $F = 50$, $in = 5$, $\boldsymbol{n} = \begin{pmatrix} 1 \\ 0 \end{pmatrix}$

15

Lesson 5
Nitsche Node-To-Node (NTN) – elmt102.f

The last considered enforcement of the non-penetration condition is the Nitsche formulation. Using the NTN discretization again, the Nitsche formulation requires additional information by its surrounding structural elements, therefore the attached trusses are presented in the following examples. The Nitsche contact element consists of the used formulation (see Section 7.3 in Part I) of three nodes where node 1 and 2 describe the contacting area whereas 2 and 3 provide the structural part for the contact formulation of the Nitsche approach, see Figure 15.1.

15.1 Implementation

Setup of tangent matrix and residual (isw = 3)

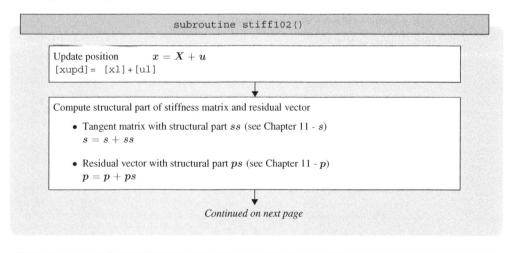

Introduction to Computational Contact Mechanics: A Geometrical Approach, First Edition.
Alexander Konyukhov and Ridvan Izi.
© 2015 John Wiley & Sons, Ltd. Published 2015 by John Wiley & Sons, Ltd.
Companion Website: www.wiley.com/go/Konyukhov

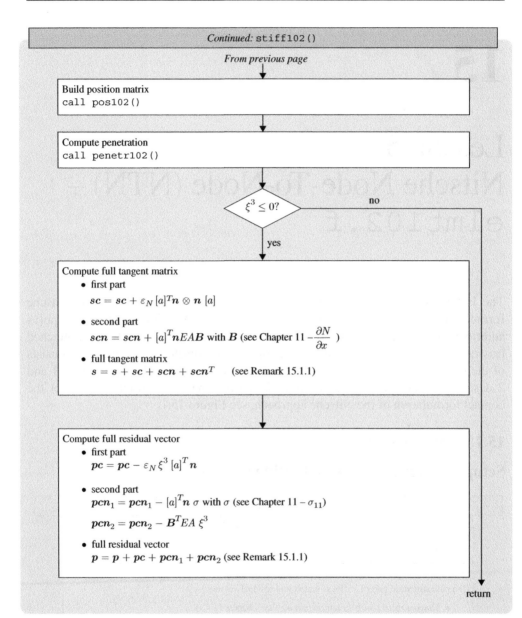

Lesson 5 Nitsche Node-To-Node (NTN) – elmt102.f

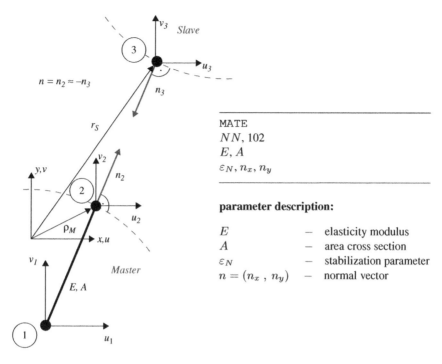

Figure 15.1 Geometry and parameters of the Nitsche NTN contact element

MATE	
NN, 102	
E, A	
ε_N, n_x, n_y	

parameter description:

E	–	elasticity modulus
A	–	area cross section
ε_N	–	stabilization parameter
$n = (n_x, n_y)$	–	normal vector

Description of subroutines

- `subroutine pos102(a)`
  Position matrix for $(r_S - \rho_M) = [a][xupd]$

$$[a] = \begin{bmatrix} -1 & 0 & 1 & 0 \\ 0 & -1 & 0 & 1 \end{bmatrix}$$

- `subroutine penetr102(xupd,n,a,penetr)`
  Penetration

$$\xi^3 = [xupd]^T [a]^T n$$

Remark 15.1.1 *Assembly of tangent matrix and residual*

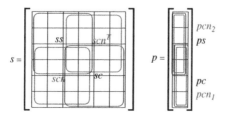

Global *FEAP*-arrays

$$[d] = \begin{bmatrix} E & A & \varepsilon_N & n_x & n_y \end{bmatrix} \qquad \text{(material parameters)}$$

$$[xl] = \begin{bmatrix} x_1 & x_2 & x_3 \\ y_1 & y_2 & y_3 \end{bmatrix} \qquad \text{(nodal coordinates)}$$

$$[ul] = \begin{bmatrix} u_1 & u_2 & u_3 \\ v_1 & v_2 & v_3 \end{bmatrix} \qquad \text{(nodal displacements)}$$

Hints for implementation

1. Entries of $[s]$ have to be symmetric as $[s]^T = [s]$ is valid.

15.2 Examples

The following examples are dealing with the NTN Nitsche formulation elmt102.f. Independent of the size of the problem setup, the Nitsche approach, in the case of truss problems, always requires three nodes in contrast to the Lagrange multiplier examples with a Schur complement where different amounts of nodes were necessary.

15.2.1 Two Trusses

The first example, as seen in the setup, deals with two trusses where the left one is subjected to a horizontal displacement of d applied in increments (in). The NTN Nitsche contact element is, hereby, defined over nodes 1–3 where the structural part of the truss is included as well as the Nitsche contact formulation.

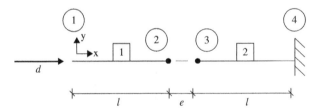

Parameters: $E = 10^3, A = 1, l = 5, e = 0.1, d = 0.12, in = 2, \varepsilon_N = 10^5, \boldsymbol{n} = \begin{pmatrix} 1 \\ 0 \end{pmatrix}$

15.2.2 Three Trusses

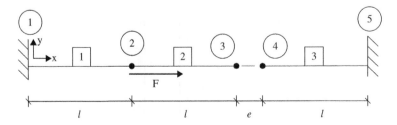

Lesson 5 Nitsche Node-To-Node (NTN) – elmt102.f

In the shown setup of three truss elements, a horizontal force F is applied in increments (in). Contact is checked between the trusses by a NTN approach. The goal is to use the implemented NTN Nitsche contact element to specify the contact condition between trusses in the 1D case (nodes 3 and 4). Define the required contact elements (elmt102.f) to solve the contact problem. Study the influence of the stabilization parameter ε_N on the value of penetration. The penetration should be independent of the stabilization parameter.

Parameters: $E = 10^3$, $A = 1$, $l = 5$, $e = 0.1$, $F = 20$, $in = 2$, $\varepsilon_N = 10^5$, $\boldsymbol{n} = \begin{pmatrix} 1 \\ 0 \end{pmatrix}$

Results of computation are given in Fig. 15.2:

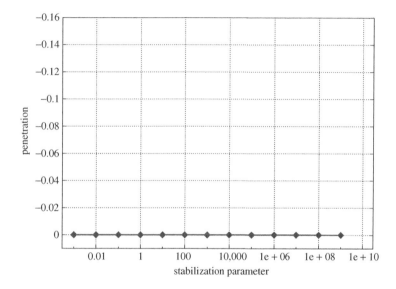

Figure 15.2 Influence of the stabilization parameter ε_N on the value of penetration

The Fig. 15.2 is illustrating the typical behavior of the Nitsche method: non-penetration condition is fulfilled exactly (penetration is zero) and is independent from the stabilization parameter.

16

Lesson 6
Node-To-Segment (NTS) – `elmt103.f`

The chapter now starts the block of various discretizations for the 2D case and even for the 3D case in Chapter 20. The contact enforcement of the non-penetration condition is, for all following elements, performed with the Penalty formulation. The first discretization presented is the Node-To-Segment (NTS), which consists in case of linear geometry approximation of the master segment of three finite element nodes, see Figure 16.1.

The NTS discretization is the most popular contact element used in FEM software to model large sliding problems with deformable bodies. Here, only deformable slave bodies with linear shape functions can be used. Higher order approximations are only valid for the master body. The theory and the corresponding assumptions for the NTS contact element are studied in Section 7.4 in Part I.

The setup of the NTS element differs compared to the Penalty NTN element in Chapter 13 by not requiring any material based input of the normal vector n. This essential attribute is realized by the closest point projection (CPP) procedure (see Section 7.4.1, Part I) in combination with the evaluation of surface geometry properties. The tangential vector is also computed from the NTS geometry and is used necessarily for the rotational part of the tangent matrix.

Special attention is given to the study of mechanical interpretation of main and rotational parts, see Remarks 7.4.5 and 7.4.6 in Chapter 7, Part I. Thus, the rotational part can be verified as an *inverted contact algorithm* for the following normal force.

16.1 Implementation

Setup of tangent matrix and residual (`isw = 3`)

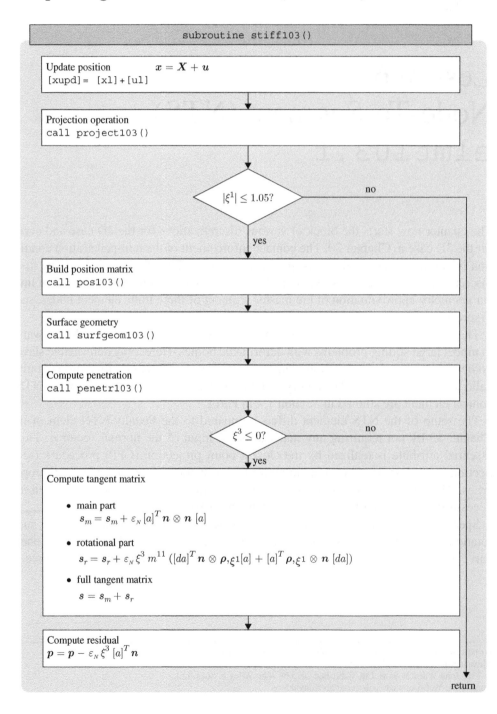

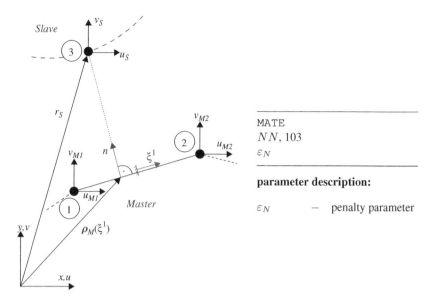

Figure 16.1 Geometry and parameters of NTS contact element

Description of subroutines

- subroutine project103(xupd,ξ^1)
  Closest point projection of $\boldsymbol{r}_S$ on $\boldsymbol{\rho}_M$

$$F(\xi^1) = \|\boldsymbol{r}_S - \boldsymbol{\rho}_M(\xi^1)\| \rightarrow \min$$

for linear NTS in 2D (see Section 7.4.1 in Part I):

$$\xi^1 = \frac{2[x_S(x_{M2}-x_{M1})+y_S(y_{M2}-y_{M1})] - x_{M2}^2 - y_{M2}^2 + x_{M1}^2 + y_{M1}^2}{(x_{M2}-x_{M1})^2 + (y_{M2}-y_{M1})^2}$$

- subroutine pos103(a,da,ξ^1)
  Position matrix for $(\boldsymbol{r}_S - \boldsymbol{\rho}_M) = [a]\,[xupd]$ and derivative
  call shape103()

$$[a] = \begin{bmatrix} -N_1 & 0 & -N_2 & 0 & 1 & 0 \\ 0 & -N_1 & 0 & -N_2 & 0 & 1 \end{bmatrix}$$

$$[da] = \frac{\partial [a]}{\partial \xi^1} = \begin{bmatrix} -N_{1,\xi^1} & 0 & -N_{2,\xi^1} & 0 & 0 & 0 \\ 0 & -N_{1,\xi^1} & 0 & -N_{2,\xi^1} & 0 & 0 \end{bmatrix}$$

- subroutine shape103(ξ^1,$\boldsymbol{N}(2,2)$)
  Shape functions and derivatives

$$[shape] = \begin{bmatrix} N_1 & N_2 \\ N_{1,\xi^1} & N_{2,\xi^1} \end{bmatrix}, \quad \text{with} \quad N_1 = \frac{1}{2}(1-\xi^1), \quad N_2 = \frac{1}{2}(1+\xi^1)$$

- subroutine surfgeom103 (xupd,da,$\boldsymbol{n}$,$\boldsymbol{\rho}_{,\xi^1}$,$m^{11}$, det$\boldsymbol{m}$)
  Surface vector
  $$\boldsymbol{\rho}_{,\xi^1} = -[da]\,[xupd]$$

Metric tensor
$$m_{11} = \boldsymbol{\rho}_{,\xi^1} \cdot \boldsymbol{\rho}_{,\xi^1}$$
$$m^{11} = \frac{1}{m_{11}}$$
$$det\boldsymbol{m} = m_{11}$$

Normal vector
$$\boldsymbol{n} = -\frac{1}{2}\frac{1}{\sqrt{det\boldsymbol{m}}}\begin{pmatrix} y_{M2}-y_{M1} \\ x_{M1}-x_{M2} \end{pmatrix}$$

- subroutine penetr103 (xupd,$\boldsymbol{n}$,a,penetr)
  Penetration
  $$\xi^3 = [xupd]^T\,[a]^T\,\boldsymbol{n}$$

Global *FEAP*-arrays

$[\mathtt{d}] = [\varepsilon_N]$ (material parameters)

$[\mathtt{xl}] = \begin{bmatrix} x_{M1} & x_{M2} & x_S \\ y_{M1} & y_{M2} & y_S \end{bmatrix}$ (nodal coordinates)

$[\mathtt{ul}] = \begin{bmatrix} u_{M1} & u_{M2} & u_S \\ v_{M1} & v_{M2} & v_S \end{bmatrix}$ (nodal displacements)

Remark 16.1.1 *The loss of continuity between linear segments leads either to multiplicity for the solution of the CPP, or to a not-allowable domain, see discussion in Section 4.2.1.2 of Part I. Numerically both lead to an oscillatory non-convergent behavior when the slave node is crossing the element boundary. The simplest way to resolve this is to allow a projection to both neighboring segments simultaneously, thus allowing overlapping segments within a certain percentage. Thus, an allowable overlapping of 5% ($|\xi^1| \leq 1.05$) is considered for the CPP procedure in order to solve this kind of problem.*

Hints for implementation

1. Using the provided Example 16.2.1, implement first the CPP procedure and carefully check the correct values of ξ^1.
2. Implement the main part of the tangent matrix only together with the residual. Contact must work already!

3. Finish the implementation with the rotational part of the tangent matrix.
4. Entries of $[s]$ have to be symmetric as $[s]^T = [s]$ is valid.
5. Residual set as $[s][ul]$ leads also to a converging computation, but with more iterations.

16.2 Examples

The examples for the NTS discretization start with simple two block setups enabling us to check the essential procedures such as the CPP procedure, the evaluation of the normal vector and setting up the main part of the tangent matrix. The third example then also includes rotations, therefore, the rotational part also becomes active.

16.2.1 Two Blocks

The standard two blocks (two elements) test is modified compared to the NTN test in Example 13.2.3, Chapter 13. One of the elements is smaller than the other and, thus, has to be chosen as the slave segment due to introduced kinematics.

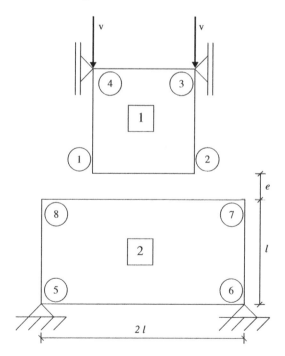

Here, a vertical displacement of v is applied in increments (in). Contact is checked between both planes by a NTS approach. The goal is to study the correct choice of the master and slave segment.

Recommended steps:

1. Supply the master segment as nodes 8–7 and the slave nodes as 1 and 2 consequently for the contact elements.
2. Consider what will happen if the master segment is chosen as nodes 1–2 and the slave points as 8 and 7.

Pay attention in all steps to the fact that reaching the maximum number of iterations for the Newton scheme means disconvergence and, in this case, an error in programming is very likely.

Parameters: $E = 10^5$, $\nu = 0.3$, $l = 1$, $v = -0.2$, $in = 20$, $e = 0.01$, $\varepsilon_N = 10^7$

16.2.2 Two Blocks – Horizontal Position

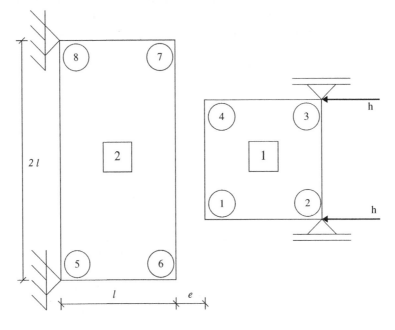

Furthermore, on the setup of two plane elements shown, a horizontal displacement of h is applied in increments (in). Contact is checked between both planes by a NTS approach. The goal is to check the correctness of the implementation regarding an arbitrary normal vector for the master segment. Define the required contact elements (elmt103.f) to solve the contact problem.

Parameters: $E = 10^5$, $\nu = 0.3$, $l = 1$, $h = -0.2$, $in = 20$, $e = 0.01$, $\varepsilon_N = 10^7$

16.2.3 Two Cantilever Beams – Large Sliding Test

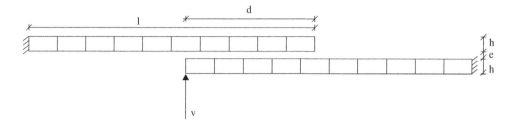

A bending problem of two beams (modeled with 10 plane elements each) enforced by contact is represented. A vertical displacement of v is applied in increments (in). The goal is to choose the correct master and slave and add necessary contact elements (elmt103.f) to solve the contact problem. Discuss the influence of the overlapping segments ($|\xi^1| \leq 1.05$, see Remark 16.1.1) on the correct kinematics and convergence.

Parameters: $E = 10^5$, $\nu = 0.3$, $l = 20$, $h = 1$, $e = 0.5$, $d = 9$, $v = 1$, $in = 10$, $\varepsilon_N = 10^6$

16.2.4 Hertz Problem

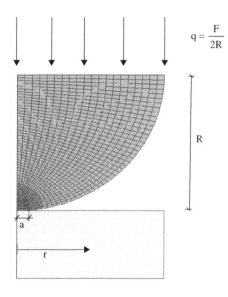

A classical Hertz contact problem (contact between a deformable cylinder and a rigid surface) (see Section 8.1 in Part I) is modeled as an axisymmetric problem with plane elements.

Goals:

1. Specify contact between the rigid surface (master) and the cylinder (slave) using NTS (elmt103.f) discretization.
2. Consider the distribution of the contact forces along the contact radius $r \in [0, a]$. a can be determined either by the:
   (a) last penetrating node, or
   (b) last positive contact force (see nodal reaction in output.dat).
3. Compare the resulting stress distribution with the analytical solution of the Hertz problem given in expression of the contact pressure as $p = \frac{2F}{\pi a^2}\sqrt{a^2 - r^2}$ where a defines the contact area radius with $a = \sqrt{\frac{4FR(1-\nu^2)}{\pi E}}$ and r represents the distance from the center of the cylinder. Perform a plot of a diagram in this case, see the result in Figure 16.2. From which side (above or below) does the numerical result converge to the analytical solution? (Correct answer is "from below".)

Parameters: $E = 10^5$, $\nu = 0$, $R = 100$, $F = 121,913.14$, $in = 5$, $\varepsilon_N = 10^8$

Results of computation compared with the analytical Hertz solution are given in Fig. 16.2:

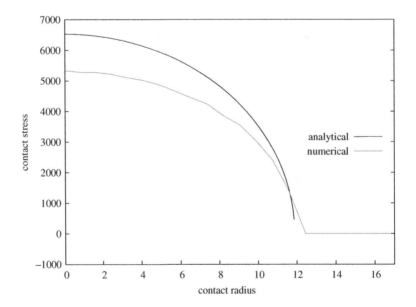

Figure 16.2 Distribution of the contact stresses along the contact radius

16.3 Inverted Contact Algorithm – Following Force

An extravagant implementation of the NTS with only rotational part and contact force N given as external loading allows us to easily interpret this case as the following normal force (always remains normal in the local coordinate system $\mathbf{N} = N\mathbf{n}$). This rather artificial structure is, however, the perfect test to check the correctness of implementation of the rotational part, the effect of which has been negligible in contact problems shown up to now.

Implementation

Setup of tangent matrix and residual (isw = 3)

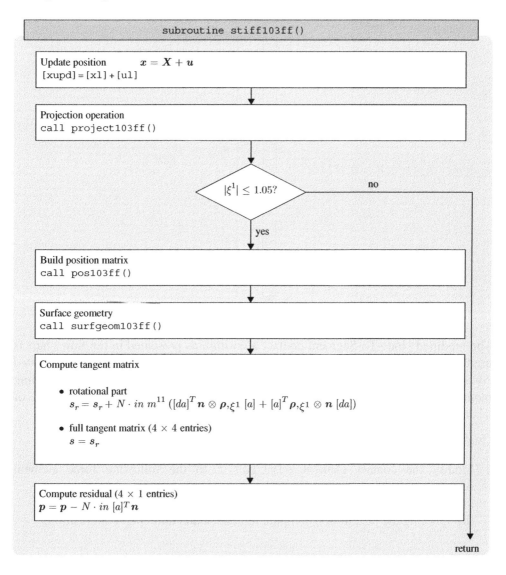

Description of subroutines
- all subroutines as in Section 16.1
- N is a single given force (F and $-F$ for two elements, respectively)
- only 4×4 and 4×1 entries for the tangent matrix and residual respectively due to the third node of the original NTS element is used only for the load location purpose.

16.3.1 Verification of the Rotational Part – A Single Following Force

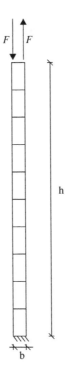

The modified Node-To-Segment contact algorithm as presented in the flow-chart is used in the current example for the modeling of following forces within *the inverted contact algorithm* as a pair of forces F in increments (in) leading to a pure bending moment. The result is correlated with a famous analytical solution for the bending of a beam into a circle, see Figure 16.3.

Goals:

1. Modify the contact element elmt103.f to the following force element elmt103ff.f by
   (a) using the incremental loading (in) on element level as in Section 14.1,
   (b) ignoring the penetration computation/check and main part computation,

(c) providing the following force (instead of ε_N),
(d) modifying the storing of tangent matrix s and residual p entries only for the first two nodes of the master segment.
2. Specify elmt103ff.f for the given example.

Parameters: $E = 10^5$, $\nu = 0$, $b = 1$, $h = 20$, $F = 9424.8$, $in = 300$

Results of computation is given in Fig. 16.3:

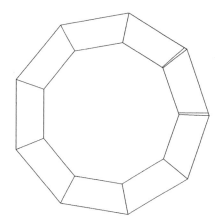

Figure 16.3   Bending of a beam into a circle

17

Lesson 7
Segment-To-Analytical-Segment (STAS) – elmt104.f

Instead of having two deformable contacting bodies, as in the case of the NTS discretization, the Segment-To-Analytical-Surface (STAS) discretization is a fast approach if one deformable body is contacting a rigid body (thus the boundary condition is given by an analytical function). The penetration is even computed in closed form if the rigid body has a simple geometry (plane, circle). Contact elements are supplied only on the deformed boundary while additional formulas defining the rigid body are necessary. The corresponding theory is given in Section 7.5 in Part I.

The following element elmt104.f is a STAS Penalty element with linear shape functions and possesses, therefore, two nodes describing the master segment. The slave is given by two possible analytically described surfaces, either by a straight line defined by two points A and B or a circle with center (x_0, y_0) and radius R, see Figure 17.1. Furthermore, the amount of integration points either Gauss or Lobatto type has to be provided. Although the flowchart contains only the Gauss quadrature, the Lobatto points are provided within the description. Since, the integration points are also the evaluation points for properties of the surface, no projection is required.

17.1 Implementation

Setup of tangent matrix and residual (`isw = 3`)

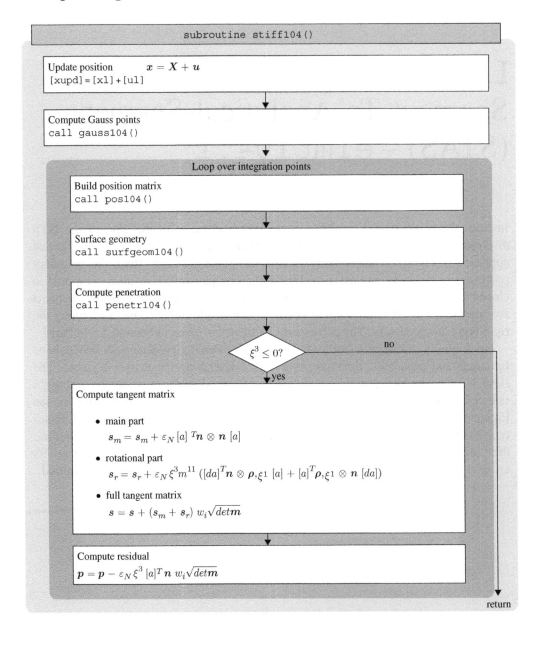

Lesson 7 Segment-To-Analytical-Segment (STAS) – elmt104.f

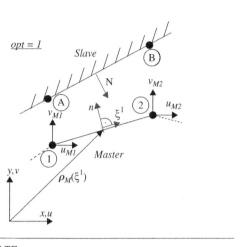
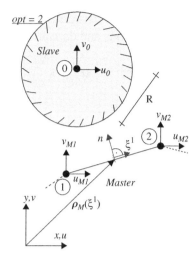

MATE
NN, 104
ε_N, ip, opt
x_A, y_A, x_B, y_B/x_0, y_0, R

parameter description:

ε_N	–	penalty parameter
ip	–	amount of integration points
opt	–	surface choice: (1) straight line (2) circle
x_A	–	x-coordinate of point A
y_A	–	y-coordinate of point A
x_B	–	x-coordinate of point B
y_B	–	y-coordinate of point B
x_0	–	x-coordinate of circle center
y_0	–	y-coordinate of circle center
R		radius

Figure 17.1 Geometry and parameters of STAS contact element: (1) contact with a rigid line; (2) contact with a rigid circle

Description of subroutines

- subroutine gauss104(ip,xi,wi)
  see Appendix A
- subroutine pos104(a,da,ξ^1)
  Position matrix for $(\boldsymbol{r}_S - \boldsymbol{\rho}_M) = [a]\,[xupd]$ and derivative
  call shape104()

$$[a] = \begin{bmatrix} -N_1 & 0 & -N_2 & 0 \\ 0 & -N_1 & 0 & -N_2 \end{bmatrix}$$

$$[da] = \frac{\partial [a]}{\partial \xi^1} = \begin{bmatrix} -N_{1,\xi^1} & 0 & -N_{2,\xi^1} & 0 \\ 0 & -N_{1,\xi^1} & 0 & -N_{2,\xi^1} \end{bmatrix}$$

- subroutine shape104 (ξ^1, $N(2,2)$)
  Shape functions and derivatives

$$[\text{shape}] = \begin{bmatrix} N_1 & N_2 \\ N_{1,\xi^1} & N_{2,\xi^1} \end{bmatrix}, \quad \text{with } N_1 = \frac{1}{2}(1-\xi^1), \quad N_2 = \frac{1}{2}(1+\xi^1)$$

- subroutine surfgeom104 (xupd, da, $\boldsymbol{n}$, $\boldsymbol{\rho}_{,\xi^1}$, m^{11}, detm)
  Surface vector

$$\boldsymbol{\rho}_{,\xi^1} = -[da]\,[xupd]$$

Metric tensor

$$m_{11} = \boldsymbol{\rho}_{,\xi^1} \cdot \boldsymbol{\rho}_{,\xi^1}$$

$$m^{11} = \frac{1}{m_{11}}$$

$$detm = m_{11}$$

Normal vector (Master)

$$\boldsymbol{n} = -\frac{1}{2}\frac{1}{\sqrt{detm}}\begin{pmatrix} y_{M2} - y_{M1} \\ x_{M1} - x_{M2} \end{pmatrix}$$

- subroutine penetr104 (d, xupd, $\boldsymbol{n}$, a, penetr)
  Master point

$$\boldsymbol{\rho} = -[a]\,[xupd]$$

Penetration including CPP procedure (closed form) for specific curves in 2D (see Section 7.5.1 in Chapter 7, Part I)

1. straight line: $\quad \xi^3 = \dfrac{(\mathbf{x}^{(A)} - \boldsymbol{\rho}) \cdot \mathbf{N}}{\mathbf{n} \cdot \mathbf{N}}$

$$\text{with } \mathbf{N} = \frac{1}{|\mathbf{x}^{(A)} - \mathbf{x}^{(B)}|}\begin{pmatrix} y_B - y_A \\ x_A - x_B \end{pmatrix}$$

2. circle: $\quad \xi^3 = \sqrt{(x_0 - \rho_x)^2 + (y_0 - \rho_y)^2} - R$

Global *FEAP*-arrays

$$[\text{d}] = \begin{bmatrix} \varepsilon_N & ip & opt & / & x_A & y_A & x_B & y_B \end{bmatrix} \quad \text{(material parameters)}$$
$$\begin{bmatrix} / & x_0 & y_0 & R \end{bmatrix}$$

$$[\text{xl}] = \begin{bmatrix} x_{M1} & x_{M2} \\ y_{M1} & y_{M2} \end{bmatrix} \qquad \text{(nodal coordinates)}$$

$$[\text{ul}] = \begin{bmatrix} u_{M1} & u_{M2} \\ v_{M1} & v_{M2} \end{bmatrix} \qquad \text{(nodal displacements)}$$

Lesson 7 Segment-To-Analytical-Segment (STAS) – elmt104.f

Hints for implementation

1. Using only one Gauss point (in the middle of the element) check the correct distance (penetration) between the contact element and the rigid surface.
2. Implement the main part of the tangent matrix only together with the residual. Contact must work!
3. Finish the implementation with the rotational part of the tangent matrix.
4. Entries of $[s]$ have to be symmetric as $[s]^T = [s]$ is valid.
5. Residual set as $[s][ul]$ leads also to a converging computation, but with more iterations.

17.2 Examples

The following examples are intended for checking the correct implementation of the STAS discretization and also to see the possible field of application starting with a horizontal rigid surface followed by an inclined surface and finishing with a rigid circular surface.

17.2.1 Block and Rigid Surface

Contact between a single structural finite element and a rigid plane aligned with the coordinate axis is the most simplest example of verifying the kinematic property of correctly implemented STAS contact elements. Due to the simplicity of the geometry the correct non-penetrability is observed even for a single integration point.

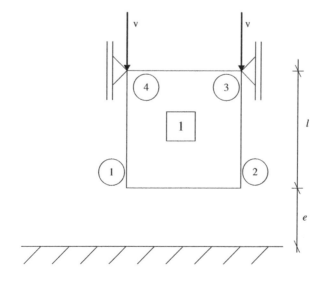

A vertical displacement of v is applied in increments (in). The goal is to analyze the influence of the amount of integration points on both the value of penetration as well as the number of iterations.

Parameters: $E = 10^5$, $\nu = 0.3$, $l = 1$, $v = -0.075$, $in = 10$, $e = 0.01$, $\varepsilon_N = 10^8$, $ip = 10$, $\mathbf{x}^{(A)} = \begin{pmatrix} 0 \\ 0 \end{pmatrix}$, $\mathbf{x}^{(B)} = \begin{pmatrix} 1 \\ 0 \end{pmatrix}$

17.2.2 Block and Inclined Rigid Surface

The following example is a more advanced test to verify the STAS contact element. The geometry represents an inclined rigid surface. The initial contact depends on the amount of integration points and on the type of integration formula (Gauss or Lobatto).

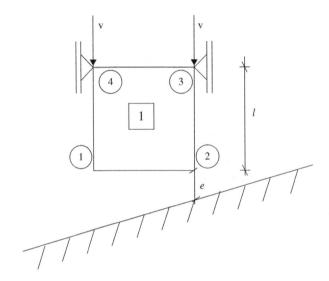

The shown setup of a square plane element over an inclined plane is used. A vertical displacement of v is applied in increments (in).
Goals:

1. Check the contact between both planes by a STAS approach.
2. Observe the correct geometry of the inclined plane if all integration points are in contact.
3. What happens if forces are applied at nodes 3 and 4 instead of displacements?

The last goal (3) here intends to study the necessity of Dirichlet boundary conditions for the solvability of structural problems in statics.

Parameters: $E = 10^5$, $\nu = 0.3$, $l = 1$, $v = -0.3$, $in = 40$, $e = 0.02$, $\varepsilon_N = 10^8$, $ip = 10$, $\mathbf{x}^{(A)} = \begin{pmatrix} 1 \\ -0.01 \end{pmatrix}$, $\mathbf{x}^{(B)} = \begin{pmatrix} 0 \\ -0.186 \end{pmatrix}$

17.2.3 Block and Inclined Rigid Surface – Different Boundary Condition

This test aims to study the influence of various boundary conditions on kinematical behaviour as well as the result of Neumann boundary conditions. Therefore, spring elements (elmt5.f) are used.

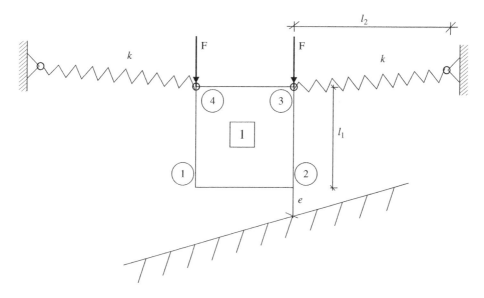

Thus, the setup shown here of a square plane attached to springs over an inclined plane is used. A vertical force of F is applied in increments (in).
Goals:

1. Check the contact between both planes by a STAS approach.
2. What happens if displacements will be applied at nodes 3 and 4 instead of forces? Discuss together with Example 17.2.2.

Parameters: $E = 10^7$, $\nu = 0.3$, $k = 8 \cdot 10^4$, $l_1 = 1$, $l_2 = 1$, $F = -50000$, $in = 50$, $e = 0.02$, $\varepsilon_N = 10^8$, $ip = 8$, $\mathbf{x}^{(A)} = \begin{pmatrix} 1 \\ -0.01 \end{pmatrix}$, $\mathbf{x}^{(B)} = \begin{pmatrix} 0 \\ -0.186 \end{pmatrix}$

17.2.4 Bending Over a Rigid Cylinder

This test is intended to verify the STAS approach for contact with a rigid circle. The contacting surface is represented during the deformation by corresponding parts (inner or outer) of the rigid circle.

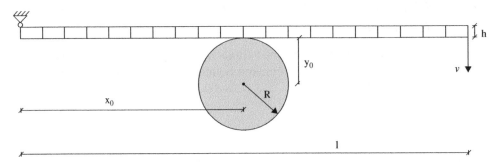

This example is representing a setup of a beam bending over a rigid circle. A vertical displacement of v is applied in increments (in).
Goals:

1. Specify the contact between the lower beam surface and a cylinder (inner part is rigid).
2. Specify the contact between the lower beam surface and a cylinder (outer part is rigid). Modifying the subroutine `penetr104` is necessary!

Parameters: $E = 10^5$, $\nu = 0.3$, $l = 40$, $h = 1$, $v = -5$, $in = 50$, $\varepsilon_N = 10^5$, $ip = 10$

Case 1: $\mathbf{x}^{(0)} = \begin{pmatrix} 20 \\ -4 \end{pmatrix}$, $R = 4$ Case 2: $\mathbf{x}^{(0)} = \begin{pmatrix} 20 \\ 48 \end{pmatrix}$, $R = 52$

17.3 Inverted Contact Algorithm – General Case of Following Forces

Using the *inverted contact algorithm* by implementation of the rotational part only and treating the normal force as externally given in the local coordinate system $\mathbf{N} = N\mathbf{n}$, we can obtain the general case of implementation for the following force. The task is a perfect test to verify the correctness of implementation for the rotational part. If the force is given at a certain point with a convective coordinate ξ_N then a single following force algorithm is obtained. If the force is approximated on a segment with corresponding shape function as $N(\xi^1) = q_1 N_1(\xi^1) + q_2 N_2(\xi^1)$ then distributed following forces algorithm is obtained. The last can be used for modeling the stress distribution during pure bending as well as for inflation problems.

Implementation

Setup of tangent matrix and residual (`isw = 3`)

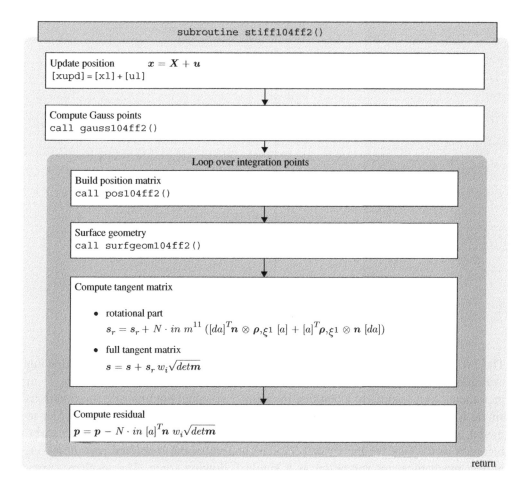

Description of subroutines

- all subroutines as in Section 17.1
- N is a given external force and computed as

$$N = \begin{cases} F \text{ and } -F & \text{for single following force} \\ q_1 N_1(\xi^1) + q_2 N_2(\xi^1) & \text{for distributed following force} \end{cases}$$

17.3.1 Verification of a Rotational Part – A Single Following Force

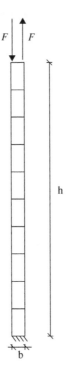

The presented Segment-To-Analytical-Segment algorithm is used in the following example for the definition of following forces applied as a pair of forces F in increments (in) leading to a bending moment. Two equal but opposite (in direction) single forces are applied to a certain point with a convective coordinate in order to model a pair of forces leading to the pure bending moment. The example can be regarded as a check for the correct implementation of the rotational part, however, with a mechanical interpretation as a single following force. The result is correlated with a famous analytical solution for the bending of a beam into a circle, see Figure 17.2, and identical to an NTS-type of single following force implementation in Figure 16.3.

Goals:

1. Modify the contact element elmt104.f to the following force element elmt104ff1.f by
   (a) using the incremental loading (in) on element level as in Section 14.1,
   (b) ignoring the loop over integration points, penetration computation/check and main part computation,
   (c) providing the projection point and the following force (instead of ε_N).

2. Specify `elmt104ff1.f` for the given example.
3. Compare with the results gained in Example 16.3.1!

Parameters: $E = 10^5, \nu = 0, b = 1, h = 20, F = 9424.8, in = 300$

Results of computation are given in Fig. 17.2:

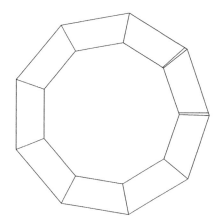

Figure 17.2 Bending of a beam into a circle: moment is modeled as a pair of single forces

17.3.2 Distributed Following Forces – Pressure

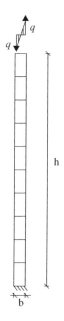

Here, the example is given as how STAS can be easily modified as an *inverted contact algorithm* in order to implement distributed normal forces. The effect of this is now used to model a linear forces distribution (pressure) as $N(\xi^1) = q_1 N_1(\xi^1) + q_2 N_2(\xi^1)$ leading to the same moment in the previous example $M = Fb = \frac{1}{2}q\frac{b}{2}\frac{4}{3}\frac{b}{2} = \frac{1}{6}qb^2$ applied in increments (in), whereas equal (but opposite in sign) nodal values of pressure are taken $q = q_2 = -q_1$.

Goals:

1. Modify the contact element elmt104.f to the pressure element elmt104ff2.f by
   (a) using the incremental loading (in) on element level as in Section 14.1,
   (b) ignoring the penetration computation/check and main part computation,
   (c) providing the amount of integration points and the linear pressure distribution (instead of ε_N).
2. Specify elmt104ff2.f for the given example.
3. Compare the current results in Figure 17.3 with the results for the single following forces in Examples shown in Figures 16.3 and 17.2. Why are the results differing (single load vs. distributed load/pressure)? (During large deformation the segment is increased in size, however, the pressure remains constant – this combination leads to the pressure distributed on the larger segment and, therefore, to the larger bending moment.)

Parameters: $E = 10^5$, $\nu = 0$, $b = 1$, $h = 20$, $q = 56548.8$, $in = 300$

Results of computation are given in Fig. 17.3:

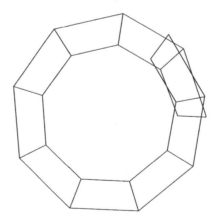

Figure 17.3 Bending of a beam into a circle: moment is modeled as exact distributed forces for pure bending

17.3.3 Inflating of a Bar

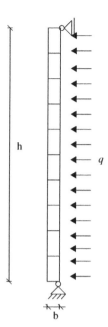

The distributed normal following forces can be used to model rising pressure. This is realized as constant nodal values of pressure $q = q_2 = q_1$. This leads to the inflation of a bar, which together with applied boundary conditions shown in the figure here deforms into a circular shaped bar (with a constant curvature) correlated with the theoretical results (see Figure 17.4). For modeling, the element elmt104ff2.f for distributed following forces is used for inflating this structure with a constant pressure distribution q in increments (in).

Parameters: $E = 10^5, \nu = 0.3, b = 0.2, h = 20, q = 2, in = 200$

Results of computation are given in Fig. 17.4:

Figure 17.4 Inflating a bar. A bar deforms into a circular shaped bar with constant curvature, which is correlated with an analytical result

18

Lesson 8
Mortar/Segment-To-Segment (STS) – `elmt105.f`

The Mortar/Segment-To-Segment (STS) type of discretization is regarded as a universal type of contact approach, see Section 7.6 in Part I. The following will be shown with numerical examples:

- By varying the number of integration points it is possible to eliminate problems with the correct choice of master and slave body as is necessary for the NTS discretization, see Remark 7.4.2 in Part I.
- The Mortar/Segment-To-Segment (STS) type discretization allows us to transfer stresses uniformly through the contacting surface, therefore, the STS approach satisfies the contact patch test.
- Any arbitrary approximation, including NURBS, high-order and iso-geometric approaches can be used.

Here, implementation is considered for linear contact segments. The STS contact element `elmt105.f`, therefore, consists of four finite element nodes, two for each surface. Since the chosen amount of integration points ip is defined on the slave surface, the CPP procedure is required for the projection of these slave points on the master segment. As before for the NTS discretization, an overlap of 5% ($|\xi^1| \leq 1.05$) is considered, see Remark 16.1.1. Also a test for verification of the rotational part is considered as *the inverted contact algorithm* with its mechanical interpretation as "inflation of the master segment caused by the slave segment".

Introduction to Computational Contact Mechanics: A Geometrical Approach, First Edition.
Alexander Konyukhov and Ridvan Izi.
© 2015 John Wiley & Sons, Ltd. Published 2015 by John Wiley & Sons, Ltd.
Companion Website: www.wiley.com/go/Konyukhov

18.1 Implementation

Setup of tangent matrix and residual (`isw = 3`)

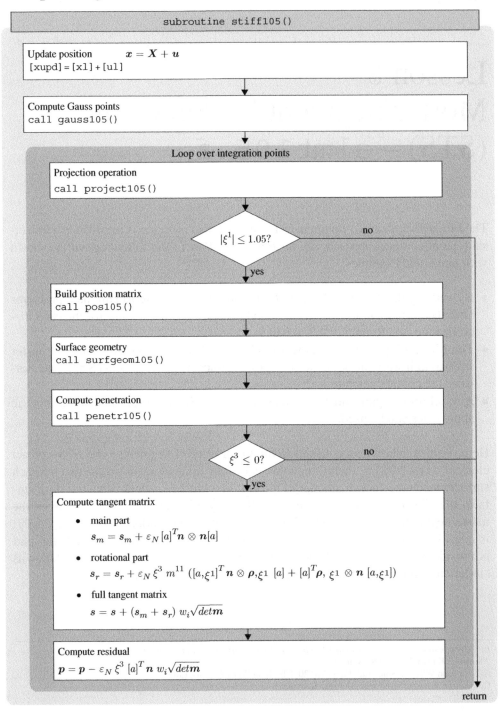

Lesson 8 Mortar/Segment-To-Segment (STS) – elmt105.f

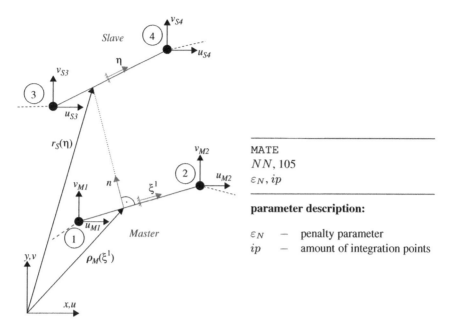

Figure 18.1 Geometry and parameters of STS contact element

Description of subroutines

- subroutine gauss105(ip,η,wi)
  see Appendix A
- subroutine project105(xupd,ξ^1,η)
  Closest point projection of r_S on ρ_M

$$F(\xi^1) = \|r_S(\eta) - \rho_M(\xi^1)\| \to min$$

for linear NTS in 2D (see Section 7.4.1 in Part I):
call shape105()

$$r_S = \begin{pmatrix} r_{Sx} \\ r_{Sy} \end{pmatrix} = N_1(\eta) \begin{pmatrix} x_{S3} \\ y_{S3} \end{pmatrix} + N_2(\eta) \begin{pmatrix} x_{S4} \\ y_{S4} \end{pmatrix}$$

$$\xi^1 = \frac{2[r_{Sx}(x_{M2} - x_{M1}) + r_{Sy}(y_{M2} - y_{M1})] - x_{M2}^2 - y_{M2}^2 + x_{M1}^2 + y_{M1}^2}{(x_{M2} - x_{M1})^2 + (y_{M2} - y_{M1})^2}$$

- subroutine pos105(a,$a_{,\xi^1}$,$a_{,\eta}$,ξ^1,η)
  Position matrix for $(r_S - \rho_M) = [a][xupd]$ and derivatives
  call shape105(ξ^1,$N(\xi^1)$)
  call shape105(η,$N(\eta)$)

$$[a] = \begin{bmatrix} -N_1(\xi^1) & 0 & -N_2(\xi^1) & 0 & N_1(\eta) & 0 & N_2(\eta) & 0 \\ 0 & -N_1(\xi^1) & 0 & -N_2(\xi^1) & 0 & N_1(\eta) & 0 & N_2(\eta) \end{bmatrix}$$

$$[a_{,\xi^1}] = \frac{\partial[a]}{\partial \xi^1} = \begin{bmatrix} -N_{1,\xi^1} & 0 & -N_{2,\xi^1} & 0 & 0 & 0 & 0 & 0 \\ 0 & -N_{1,\xi^1} & 0 & -N_{2,\xi^1} & 0 & 0 & 0 & 0 \end{bmatrix}$$

$$[a_{,\eta}] = \frac{\partial[a]}{\partial \eta} = \begin{bmatrix} 0 & 0 & 0 & 0 & N_{1,\eta} & 0 & N_{2,\eta} & 0 \\ 0 & 0 & 0 & 0 & 0 & N_{1,\eta} & 0 & N_{2,\eta} \end{bmatrix}$$

- subroutine shape105($x, N(2,2)$)
  Shape functions and derivatives

$$[\text{shape}] = \begin{bmatrix} N_1 & N_2 \\ N_{1,x} & N_{2,x} \end{bmatrix}, \text{ with } N_1 = \frac{1}{2}(1-x),\, N_2 = \frac{1}{2}(1+x)$$

- subroutine surfgeom105(xupd, $a_{,\xi^1}, a_{,\eta}, n, \rho_{,\xi^1}, m^{11}, detm$)
  Surface vector (Slave)

$$\boldsymbol{r}_{,\eta} = -[a_{,\eta}][xupd]$$

Metric tensor (Slave)

$$detm = \boldsymbol{r}_{,\eta} \cdot \boldsymbol{r}_{,\eta}$$

Surface vector (Master)

$$\boldsymbol{\rho}_{,\xi^1} = -[a_{,\xi^1}][xupd]$$

Metric tensor (Master)

$$m_{11} = \boldsymbol{\rho}_{,\xi^1} \cdot \boldsymbol{\rho}_{,\xi^1}$$

$$m^{11} = \frac{1}{m_{11}}$$

Normal vector (Master)

$$\boldsymbol{n} = -\frac{1}{2\sqrt{\boldsymbol{\rho}_{,\xi^1} \cdot \boldsymbol{\rho}_{,\xi^1}}} \begin{pmatrix} y_{M2} - y_{M1} \\ x_{M1} - x_{M2} \end{pmatrix}$$

- subroutine penetr105(xupd, n, a, penetr)
  Penetration

$$\xi^3 = [xupd]^T [a]^T \boldsymbol{n}$$

Global *FEAP*-arrays

$$[\text{d}] = \begin{bmatrix} \varepsilon_N & ip \end{bmatrix} \qquad \text{(material parameters)}$$

$$[\text{xl}] = \begin{bmatrix} x_{M1} & x_{M2} & x_{S3} & x_{S4} \\ y_{M1} & y_{M2} & y_{S3} & y_{S4} \end{bmatrix} \qquad \text{(nodal coordinates)}$$

$$[\text{ul}] = \begin{bmatrix} u_{M1} & u_{M2} & u_{S3} & u_{S4} \\ v_{M1} & v_{M2} & v_{S3} & v_{S4} \end{bmatrix} \qquad \text{(nodal displacements)}$$

Hints for implementation

1. Using only one Gauss point (in the middle of the element) check the correct distance (penetration) between the contact segments.
2. Implement the main part of the tangent matrix only together with the residual. Contact must work!
3. Finish the implementation with the rotational part of the tangent matrix.
4. Entries of $[s]$ have to be symmetric as $[s]^T = [s]$ is valid.
5. Residual set as $[s][ul]$ leads also to a converging computation, but with more iterations.

18.2 Examples

The STS discretization will be first applied to examples with the same setup as for the NTS and STAS discretization, in order to compare these. The third example is the contact patch test, which in specific is applicable to the STS discretization.

18.2.1 Two Blocks

As has been investigated in Example 16.2.1, the correct choice of the master segment is important for the NTS approach. In case of the STS discretization, the master and slave segment can be selected arbitrarily, however, the number of integration points should be taken into account as an important parameter.

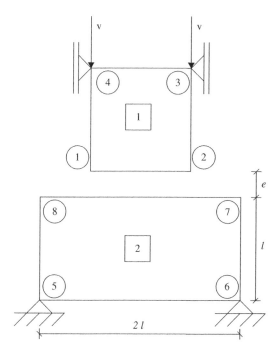

A vertical displacement of v is applied in increments (in).
Goals:

1. Contact is checked between both planes by a STS approach. Define the required contact elements (elmt105.f) to solve the problem.
2. Supply master as the lower body and slave as the upper body. Compare the results (displacements) with the NTS approach.
3. Study the influence of the number of integration points $ip = 1...10$.
4. Change the order so that the master is the upper body and the slave is lower body. Study the influence of the number of integration points.
5. Modify the mesh of the lower body by splitting into two plane elements. Do the displacements change? Study the influence of the number of integration points.

Parameters: $E = 10^5$, $\nu = 0.3$, $l = 1$, $v = -0.1$, $in = 10$, $e = 0.01$, $\varepsilon_N = 10^7$, $ip = 10$

18.2.2 Block and Inclined Rigid Surface – Different Boundary Condition

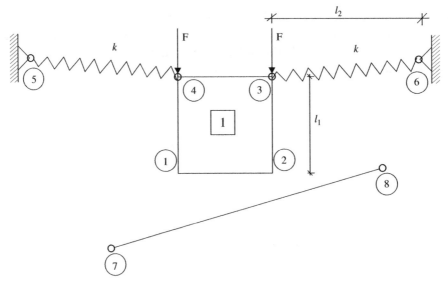

The second example is the setup shown here of a square plane attached to springs over an inclined plane, the same as in Example 17.2.3. A vertical force of F is applied in increments (in). Contact is now checked between both planes by a STS approach.
Goals:

1. Supply the master for the lower body and the slave for the upper body.
2. Consider the solution with NTS contact elements and a STS contact element with only two integration points.

3. Study the influence of the amount of integration points on the solution ($ip = 1...10$). How many integration points are enough to achieve reasonably correct results?
4. Change the choice of master and slave within STS and study goal (3) again.

Parameters: $E = 10^6$, $\nu = 0.3$, $k = 8 \cdot 10^4$, $l_1 = 1$, $l_2 = 1$, $F = -80,000$, $in = 80$
$\varepsilon_N = 10^8$, $ip = 5$

18.2.3 Contact Patch Test

The last STS example is the contact patch test. The contact patch test serves to check if homogenous stresses (according to the available simplest analytical solution) are transferred through the contact surface uniformly. The patch test is said to be fully (unconditionally) satisfied if the stresses are transferred uniformly independently of the choice of master and slave segment. Otherwise either "no fulfillment" or "conditional fulfillment" (e.g., proper choice of master and slave) can be stated for a certain contact approach. Displacements on the contact segment can be used as the simplest measure instead of stresses.

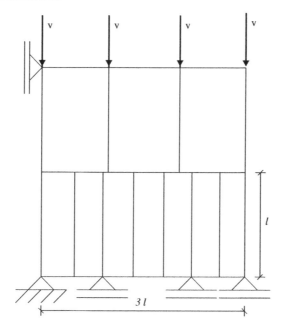

The setup shown here of two plane element blocks is used. The upper block is discretized with 3 elements, the lower one with 7. A vertical displacement of v is applied in increments (in) on the upper surface. Use both (NTS and STS) approaches to solve the problem. What differences can be found? What changes while varying the amount of integration points ($ip = 1 \ldots 10$)?

Parameters: $E = 10^5$, $\nu = 0.3$, $l = 1$, $v = -0.1$, $in = 20$, $ip = 10$, $\varepsilon_N = 10^7$

18.3 Inverted Contact Algorithm – Following Force

Again, similar to both the NTS (in Section 16.3) and to the STAS (in Section 17.3) algorithms, it is possible to implement "the inverted contact algorithm" by considering the rotational part only and contact force N given as external loading. One should keep in mind, however, that the following force is given now as a projection of the slave Gauss point in the coordinate system of the master $\mathbf{N} = N(\xi^1)\mathbf{n}(\xi^1)$. Thus, the normal force is "the following force" only for the master side – on the slave side this force acts with the same value, but in the direction opposite to the master normal.

Implementation

Setup of the tangent matrix and residual (isw = 3)

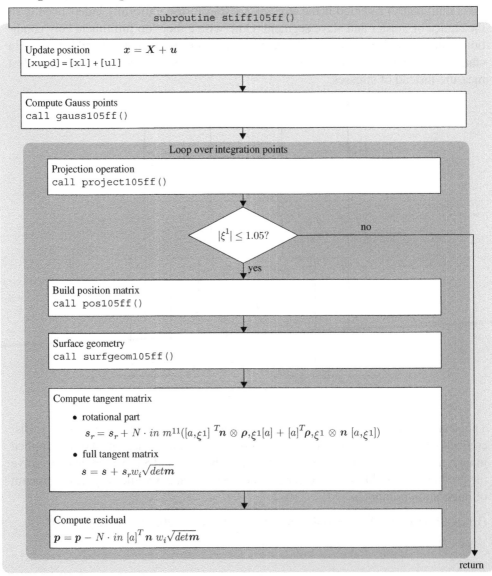

18.3.1 Verification of the Rotational Part – Pressure on the Master Side

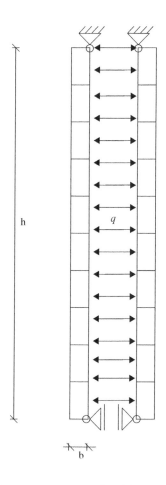

The modified Segment-To-Segment contact algorithm, as presented in the flowchart, is used in the current example for the modeling of following forces within *the inverted contact algorithm* acting on the master side. The forces acting on the slave side are normal to the master side, but not to a slave side! However, the result looks like a pressure load between both master and slave sides, this is *by no means a pressure inflating a gap, but only pressure acting on the master side*. The example is, however, a perfect test for the correct implementation of the rotational part.

Goals:

1. Modify the contact element elmt105.f to the following force element elmt105ff.f by
   (a) using the incremental loading (in) on the element level as in Section 14.1,

(b) ignoring penetration computation/check and main part computation, whilst keeping, however, the projection check active,
(c) providing the amount of integration points and the constant pressure distribution (instead of ε_N) as $N = qN_1(\eta^1) + qN_2(\eta^1)$ on the slave part.

Parameters: $E = 10^5$, $\nu = 0.3$, $b = 0.2$, $h = 20$, $q = 0.3$, $in = 300$

Results of computation are given in Fig. 18.2:

Figure 18.2 Within the STS-type inverted contact algorithm the forces follow the master normal only. The following master segment forces act on the slave segment. This is by no means inflating pressure for both sides, but a test for the rotational part!

19

Lesson 9
Higher Order Mortar/STS – `elmt106.f`

The Mortar or STS contact approach is necessary for higher order finite elements, NURBS or iso-geometric finite elements. Examples here are the Lagrange, Lobatto and Bezier class of shape functions with quadratic order for the corresponding contacting segments. In order to implement the STS discretization for any order and type of approximation the corresponding position matrix $[a]$ has to be modified with regard to the class of chosen shape functions on the master $N_i(\xi)$ and on the slave $N_i(\eta)$ segments. Thus, the closed form of the CPP procedure applied for linear segments in 2D is no longer valid and has to be modified into the full iterative solution scheme using Newton's method (see Section 4.2.2 in Part I) instead of the closed form solution used previously for the linear segments. An overlap of 5% ($|\xi^1| \leq 1.05$) as in case of linear approximations is, for these higher approximations, not necessarily required but can still be used, see Section 4.2.1.2 in Part I and Remark 16.1.1.

In the case of `elmt106.f`, six element nodes/knots are defined, three for the master and three for the slave segment, as can be seen in Figure 19.1 for the quadratic Lagrange type. The other two classes of shape functions are described in more detail in Solin and Segeth (2004). Due to higher order approximation the master surface possesses a curvature that leads to additional curvature part entries within the tangent matrix.

19.1 Implementation

Setup of tangent matrix and residual (`isw = 3`)

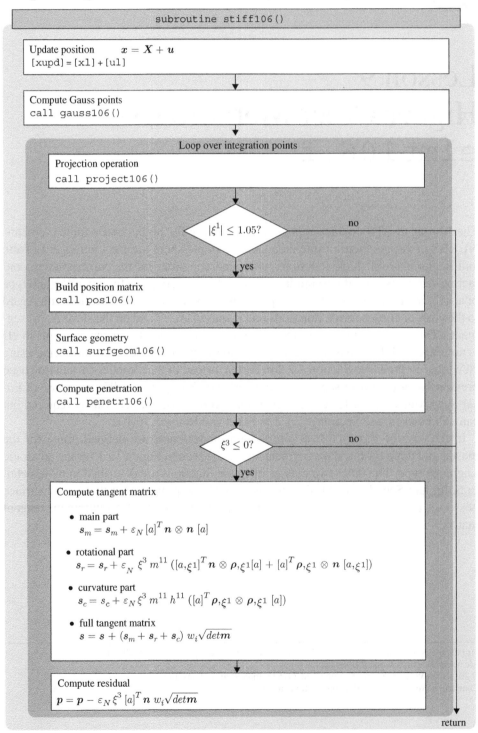

Lesson 9 Higher Order Mortar/STS – `elmt106.f`

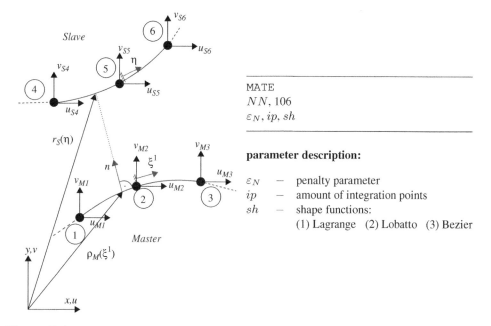

MATE
NN, 106
ε_N, ip, sh

parameter description:

ε_N –   penalty parameter
ip –   amount of integration points
sh –   shape functions:
         (1) Lagrange   (2) Lobatto   (3) Bezier

Figure 19.1 Geometry and parameters of STS contact element with higher order approximations

Description of subroutines

- `subroutine gauss106(ip,`η`,wi)`
  see Appendix A
- `subroutine project106(xupd,`ξ^1,η,d`)`
  Closest point projection of r_S on ρ_M

$$F(\xi^1) = \|r_S(\eta) - \rho_M(\xi^1)\| \to \min$$

Newton's method with $\xi^1_{(n+1)} = \xi^1_{(n)} + \Delta\xi^1_{(n)}$ and $\xi^1_{(0)} = 0$
where

$$\Delta\xi^1_{(n)} = -(F'')^{-1}_{(n)} F'_{(n)} \quad \text{with} \quad F(\xi^1) = \frac{1}{2}(r_S - \rho_M) \cdot (r_S - \rho_M)$$

$$F' = (r_S - \rho_M(\xi^1)) \cdot \frac{\partial \rho_M}{\partial \xi^1}$$

$$F'' = (r_S - \rho_M(\xi^1)) \cdot \frac{\partial^2 \rho_M}{\partial^2 \xi^1} - \frac{\partial \rho_M}{\partial \xi^1} \cdot \frac{\partial \rho_M}{\partial \xi^1}$$

$$\Delta\xi^1_{(n)} = -\frac{(r_S - \rho_M) \cdot \rho_{,\xi^1}}{(r_S - \rho_M) \cdot \rho_{,\xi^1\xi^1} - \rho_{,\xi^1} \cdot \rho_{,\xi^1}}$$

```
call pos106()
call surfgeom106()
```

$$\Delta \xi_{(n)}^1 = -\frac{f}{e - \boldsymbol{\rho}_{,\xi^1} \cdot \boldsymbol{\rho}_{,\xi^1}} \quad \text{with } f = \boldsymbol{\rho}_{,\xi^1}[a][xupd], \; e = \boldsymbol{\rho}_{,\xi^1\xi^1}[a][xupd]$$

- subroutine pos106($a, a_{,\xi^1}, a_{,\xi^1\xi^1}, a_{,\eta}, \xi^1, \eta, d$)
  Position matrix for $(r_S - \rho_M) = [a][xupd]$ and derivatives
  call shape106($\xi^1, N(\xi^1), d$)
  call shape106($\eta, N(\eta), d$)

$$[a] = \begin{bmatrix} -N_1(\xi^1) & 0 & -N_2(\xi^1) & 0 & \cdots & N_1(\eta) & 0 & N_2(\eta) & 0 & N_3(\eta) & 0 \\ 0 & -N_1(\xi^1) & 0 & -N_2(\xi^1) & \cdots & 0 & N_1(\eta) & 0 & N_2(\eta) & 0 & N_3(\eta) \end{bmatrix}$$

$$[a_{,\xi^1}] = \frac{\partial[a]}{\partial \xi^1} = \begin{bmatrix} -N_{1,\xi^1} & 0 & -N_{2,\xi^1} & 0 & -N_{3,\xi^1} & 0 & 0 & \cdots & 0 & 0 \\ 0 & -N_{1,\xi^1} & 0 & -N_{2,\xi^1} & 0 & -N_{3,\xi^1} & 0 & \cdots & 0 & 0 \end{bmatrix}$$

$$[a_{,\xi^1\xi^1}] = \frac{\partial^2[a]}{\partial \xi^{1\,2}} = \begin{bmatrix} -N_{1,\xi^1\xi^1} & 0 & -N_{2,\xi^1\xi^1} & 0 & -N_{3,\xi^1\xi^1} & 0 & 0 & \cdots & 0 & 0 \\ 0 & -N_{1,\xi^1\xi^1} & 0 & -N_{2,\xi^1\xi^1} & 0 & -N_{3,\xi^1\xi^1} & 0 & \cdots & 0 & 0 \end{bmatrix}$$

$$[a_{,\eta}] = \frac{\partial[a]}{\partial \eta} = \begin{bmatrix} 0 & 0 & \cdots & 0 & N_{1,\eta} & 0 & N_{2,\eta} & 0 & N_{3,\eta} & 0 \\ 0 & 0 & \cdots & 0 & 0 & N_{1,\eta} & 0 & N_{2,\eta} & 0 & N_{3,\eta} \end{bmatrix}$$

- subroutine shape106($x, N(3,3), d$)
  Shape functions and derivatives

$$[\text{shape}] = \begin{bmatrix} N_1 & N_2 & N_3 \\ N_{1,x} & N_{2,x} & N_{3,x} \\ N_{1,xx} & N_{2,xx} & N_{3,xx} \end{bmatrix},$$

with

1. Lagrange: $N_1 = \frac{1}{2}x(x-1),\; N_2 = 1 - x^2,\; N_3 = \frac{1}{2}x(x+1)$

2. Lobatto: $N_1 = \frac{1}{2}(1-x),\; N_2 = \frac{1}{2}\sqrt{\frac{3}{2}}(x^2-1),\; N_3 = \frac{1}{2}(1+x)$

3. Bezier: $N_1 = \frac{1}{4}(1-x)^2,\; N_2 = \frac{1}{2}(1-x)(1+x),\; N_3 = \frac{1}{4}(1+x)^2$

- subroutine surfgeom106(xupd, $a_{,\xi^1}, a_{,\xi^1\xi^1}, a_{,\eta}, \boldsymbol{n}, \boldsymbol{\rho}_{,\xi^1}, \boldsymbol{\rho}_{,\xi^1\xi^1}, m^{11}, h^{11}, \det\boldsymbol{m}$)
  Surface vector (Slave)

$$\boldsymbol{r}_{,\eta} = -[a_{,\eta}][xupd]$$

Metric tensor (Slave)

$$\det\boldsymbol{m} = \boldsymbol{r}_{,\eta} \cdot \boldsymbol{r}_{,\eta}$$

Surface vector (Master)

$$\boldsymbol{\rho}_{,\xi^1} = -[a_{,\xi^1}][xupd]$$

Metric tensor (Master)

$$m_{11} = \boldsymbol{\rho}_{,\xi^1} \cdot \boldsymbol{\rho}_{,\xi^1}$$

$$m^{11} = \frac{1}{m_{11}}$$

Normal vector (Master)

$$n = \frac{1}{\sqrt{\rho_{,\xi^1} \cdot \rho_{,\xi^1}}} \begin{pmatrix} -\rho_{,\xi^1}(2) \\ \rho_{,\xi^1}(1) \end{pmatrix}$$

Second derivative of $\rho(\xi^1)$

$$\rho_{,\xi^1\xi^1} = -[a_{,\xi^1\xi^1}][xupd]$$

Curvature tensor

$$h^{11} = \frac{\rho_{,\xi^1\xi^1} \cdot n}{(\rho_{,\xi^1} \cdot \rho_{,\xi^1})^2}$$

- `subroutine penetr106(xupd,n,a,penetr)`
  Penetration

$$\xi^3 = [xupd]^T [a]^T n$$

Global *FEAP*-arrays

$[d] = \begin{bmatrix} \varepsilon_N & ip & sh \end{bmatrix}$ (material parameters)

$[x1] = \begin{bmatrix} x_{M1} & x_{M2} & x_{M3} & x_{S4} & x_{S5} & x_{S6} \\ y_{M1} & y_{M2} & y_{M3} & y_{S4} & y_{S5} & y_{S6} \end{bmatrix}$ (nodal coordinates)

$[u1] = \begin{bmatrix} u_{M1} & u_{M2} & u_{M3} & u_{S4} & u_{S5} & u_{S6} \\ v_{M1} & v_{M2} & v_{M3} & v_{S4} & v_{S5} & v_{S6} \end{bmatrix}$ (nodal displacements)

Hints for implementation

1. Using the provided Example 19.2.1, implement first the CPP procedure and carefully check the correct values of ξ^1.
2. Using only one Gauss point (in the middle of the element) check the correct distance (penetration) between the contact segments.
3. Implement the main part of the tangent matrix only together with the residual. Contact must work!
4. Finish the implementation with the rotational and curvature part of the tangent matrix.
5. Entries of $[s]$ have to be symmetric as $[s]^T = [s]$ is valid.
6. Residual set as $[s][ul]$ leads also to a converging computation, but with more iterations.

19.2 Examples

The examples given here for STS discretization with different quadratic shape functions classes are comparable with the examples of the linear STS approach in Section 18.2. In order to use the same shape functions for the structural plane elements also, the provided finite element code already provides a plane element of these quadratic shape function classes (`elmt6.f`). Thus, the structural and contact surface description are on top of each other. Although the geometry is described differently regarding the setup of nodes/knots for the different classes of shape functions, the attempt is to establish the same curved geometry using these different classes.

19.2.1 Two Blocks

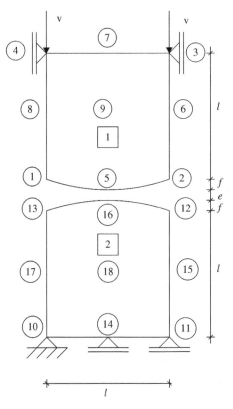

The setup shown here of two plane elements is used. A vertical displacement of v is applied in increments (in).
Goals:

1. Contact is checked between both planes by a STS approach. Define the required contact elements (elmt106.f) to solve the problem.
2. Supply master as the lower body and slave as the upper body. Compare the results (displacements) for the different shape function classes Lagrange, Lobatto and Bezier.
3. Study the influence of the number of integration points $ip = 1...10$.
4. Plot the initial and final configuration for each shape function class. What difference can be seen for the Lobatto class?

Parameters: $E = 10^5$, $\nu = 0.3$, $l = 2$, $v = -0.3$, $in = 30$, $e = 0.015$, $f = 0.2$, $\varepsilon_N = 10^7$, $ip = 10$

Results of computation for Lagrange, Lobatto and Bezier shape functions are given in Fig. 19.2:

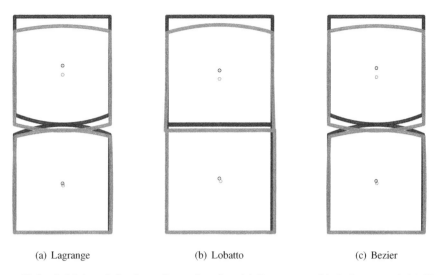

Figure 19.2 Initial and final configuration for (a) Lagrange, (b) Lobatto and (c) Bezier classes of shape functions

19.2.2 Block and Inclined Rigid Surface – Different Boundary Condition

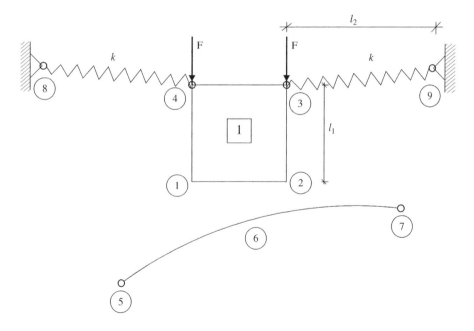

Next, the setup shown here of a square plane attached to springs over an inclined plane is used. A vertical force of F is applied in increments (in). Contact is now checked between both planes by a STS approach.

Goals:

1. Supply master for the lower body and slave for the upper body.
2. Plot the initial and final configuration for each shape function class. What difference can be seen?

Parameters: $E = 10^5$, $\nu = 0.3$, $k = 8 \cdot 10^4$, $l_1 = 1$, $l_2 = 1$, $F = -70,000$, $in = 70$, $\varepsilon_N = 10^8$, $ip = 5$

Results of computation for Lagrange, Lobatto and Bezier shape functions are given in Fig. 19.3:

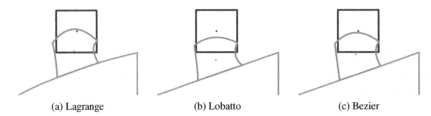

(a) Lagrange   (b) Lobatto   (c) Bezier

Figure 19.3 Initial and final configuration for (a) Lagrange, (b) Lobatto and (c) Bezier classes of shape functions

20

Lesson 10
3D Node-To-Segment (NTS) – `elmt107.f`

The last chapter within the block of various discretizations deals with the Node-To-Segment (NTS) discretization as in Chapter 16, but for problems in 3D. This Chapter is devoted to show particular implementation features arising for 3D problems such as the requirements of a surface metric tensor in co- and contravariant form, see the sections on differential geometry of surfaces, such as Sections 3.3 and 3.4 in Part I; implementation of the full CPP procedure for two convective coordinates ξ^1, ξ^2, see Section 4.2.3 in Part I and consideration of 3D contact kinematics, see Section 4.3.2 in Part I. The implementation is intentionally shown for the most simplest case of NTS with linear shape functions and penalty enforcement, see Figure 20.1. Reaching this stage of the programming, the reader is prompted to combine various contact approaches (STS, STAS) with various enforcement methods (Lagrange, Penalty, Nitsche, Augmented Lagrange) for 3D non-frictional problems based on his/her personal research needs.

The 3D NTS element `elmt107.f` consists of four nodes on the master surface and one additional node on the slave surface. As mentioned in Remark 16.1.1, the CPP procedure for both ξ^1 and ξ^2 will be handled here again with an overlap of 5%. Furthermore, the rotational part of the tangent matrix has more terms due to the matrix character of the metric tensor.

Introduction to Computational Contact Mechanics: A Geometrical Approach, First Edition.
Alexander Konyukhov and Ridvan Izi.
© 2015 John Wiley & Sons, Ltd. Published 2015 by John Wiley & Sons, Ltd.
Companion Website: www.wiley.com/go/Konyukhov

20.1 Implementation

Setup of tangent matrix and residual (isw = 3)

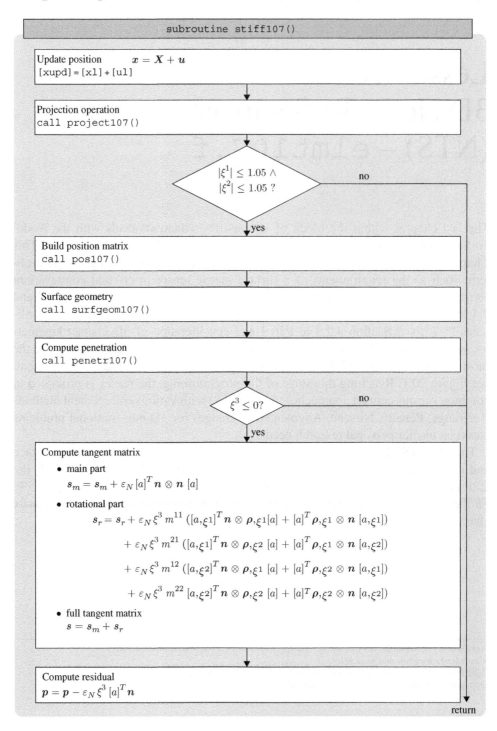

Lesson 10 3D Node-To-Segment (NTS) – elmt107.f

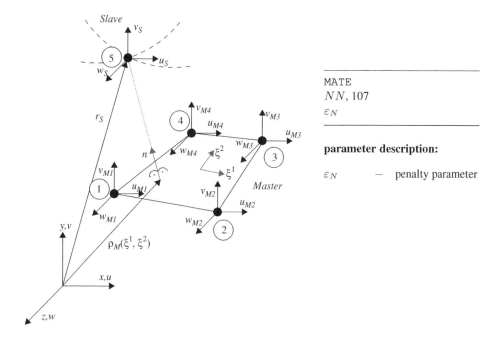

MATE
NN, 107
ε_N

parameter description:

ε_N — penalty parameter

Figure 20.1 Geometry and parameters of the 3D NTS contact element

Description of subroutines

- subroutine project107(xupd,ξ^1,ξ^2)
  Closest point projection of r_S on ρ_M

$$F(\xi^1,\xi^2) = \|r_S - \rho_M(\xi^1,\xi^2)\| \to min$$

Newton's method with $\boldsymbol{\xi}_{(n+1)} = \boldsymbol{\xi}_{(n)} + \Delta\boldsymbol{\xi}_{(n)}$ and $\boldsymbol{\xi}_{(0)} = \begin{pmatrix} 0 \\ 0 \end{pmatrix}$

where

$$\Delta\boldsymbol{\xi}_{(n)} = -(F'')^{-1}_{(n)} F'_{(n)} \quad \text{with} \quad F(\xi^1,\xi^2) = \frac{1}{2}(r_S - \rho_M) \cdot (r_S - \rho_M)$$

$$F' = -\begin{bmatrix} \rho_{,1} \cdot (r_S - \rho_M) \\ \rho_{,2} \cdot (r_S - \rho_M) \end{bmatrix} \quad \text{with} \quad \rho_{,i} = \frac{\partial \rho}{\partial \xi^i}$$

$$F'' = -\begin{bmatrix} \rho_{,1} \cdot \rho_{,1} - \rho_{,11} \cdot (r_S - \rho_M) & \rho_{,1} \cdot \rho_{,2} - \rho_{,12} \cdot (r_S - \rho_M) \\ \rho_{,2} \cdot \rho_{,1} - \rho_{,21} \cdot (r_S - \rho_M) & \rho_{,2} \cdot \rho_{,2} - \rho_{,22} \cdot (r_S - \rho_M) \end{bmatrix}$$

$$\Delta\boldsymbol{\xi}_{(n)} = \frac{1}{\det F''} \begin{bmatrix} \rho_{,2} \cdot \rho_{,2} - \rho_{,22} \cdot (r_S - \rho_M) & \rho_{,12} \cdot (r_S - \rho_M) - \rho_{,1} \cdot \rho_{,2} \\ \rho_{,21} \cdot (r_S - \rho_M) - \rho_{,2} \cdot \rho_{,1} & \rho_{,1} \cdot \rho_{,1} - \rho_{,11} \cdot (r_S - \rho_M) \end{bmatrix} \begin{bmatrix} \rho_{,1} \cdot (r_S - \rho_M) \\ \rho_{,2} \cdot (r_S - \rho_M) \end{bmatrix}$$

for linear shape functions

$$\Rightarrow \rho_{,ii} = 0 \text{ and } \rho_{,12} = \rho_{,21} = \frac{1}{4}(x^1 - x^2 + x^3 - x^4)$$

```
call shape107()
call pos107()
call surfgeom107()
```

$$\Delta\xi_{(n)} = \begin{pmatrix} \Delta\xi^1_{(n)} \\ \Delta\xi^2_{(n)} \end{pmatrix} =$$

$$= \frac{1}{detm - e^2 + 2em_{12}} \begin{bmatrix} m_{22} & e - m_{12} \\ e - m_{21} & m_{11} \end{bmatrix} \begin{bmatrix} f_1 \\ f_2 \end{bmatrix}$$

with $f_i = \rho_{,i}[a][xupd]$, $e = \rho_{,12}[a][xupd]$

- `subroutine pos107(a,`$a_{,\xi^1}, a_{,\xi^2}, \xi^1, \xi^2$`)`

 Position matrix for $(r_S - \rho_M) = [a][xupd]$ and derivatives
 ```
 call shape107()
 ```

$$[a] = \begin{bmatrix} -N_1 & 0 & 0 & -N_2 & 0 & 0 & -N_3 & 0 & 0 & -N_4 & 0 & 0 & 1 & 0 & 0 \\ 0 & -N_1 & 0 & 0 & -N_2 & 0 & 0 & -N_3 & 0 & 0 & -N_4 & 0 & 0 & 1 & 0 \\ 0 & 0 & -N_1 & 0 & 0 & -N_2 & 0 & 0 & -N_3 & 0 & 0 & -N_4 & 0 & 0 & 1 \end{bmatrix}$$

$$[a_{,\xi^1}] = \frac{\partial[a]}{\partial\xi^1}, \quad [a_{,\xi^2}] = \frac{\partial[a]}{\partial\xi^2}$$

- `subroutine shape107(`ξ^1, ξ^2`, shape)`

 Shape functions

$$[shape] = \begin{bmatrix} N_1 & N_2 & N_3 & N_4 \\ N_{1,\xi^1} & N_{2,\xi^1} & N_{3,\xi^1} & N_{4,\xi^1} \\ N_{1,\xi^2} & N_{2,\xi^2} & N_{3,\xi^2} & N_{4,\xi^2} \end{bmatrix}$$

with

$$N_1 = \frac{1}{4}(1-\xi^1)(1-\xi^2), \quad N_2 = \frac{1}{4}(1+\xi^1)(1-\xi^2)$$

$$N_3 = \frac{1}{4}(1+\xi^1)(1+\xi^2), \quad N_4 = \frac{1}{4}(1-\xi^1)(1+\xi^2)$$

- `subroutine surfgeom107(`$xupd, a_{,\xi^1}, a_{,\xi^2}, n, \rho_{,\xi^1}, \rho_{,\xi^2}, m, detm, m^{-1}$`)`

 Surface vectors

$$\rho_{,\xi^1} = -[a_{,\xi^1}][xupd]$$

$$\rho_{,\xi^2} = -[a_{,\xi^2}][xupd]$$

Metric tensor

$$m = \begin{bmatrix} \rho_{,\xi^1} \cdot \rho_{,\xi^1} & \rho_{,\xi^1} \cdot \rho_{,\xi^2} \\ \rho_{,\xi^2} \cdot \rho_{,\xi^1} & \rho_{,\xi^2} \cdot \rho_{,\xi^2} \end{bmatrix}$$

Lesson 10 3D Node-To-Segment (NTS) – elmt107.f

$$det\boldsymbol{m} = m_{11} \cdot m_{22} - m_{21} \cdot m_{12}$$

$$\boldsymbol{m}^{-1} = \frac{1}{det\boldsymbol{m}} \begin{bmatrix} m_{22} & -m_{12} \\ -m_{21} & m_{11} \end{bmatrix}$$

Normal vector

$$\boldsymbol{n} = \frac{\boldsymbol{\rho}_{,\xi_1} \times \boldsymbol{\rho}_{,\xi_2}}{\sqrt{det\boldsymbol{m}}}$$

- subroutine penetr107(xupd,n,a,penetr)
  Penetration

$$\xi^3 = [xupd]^T [a]^T \boldsymbol{n}$$

Global *FEAP*-arrays

$[d] = [\varepsilon_N]$ (material parameters)

$$[\mathrm{x}1] = \begin{bmatrix} x_{M1} & x_{M2} & x_{M3} & x_{M4} & x_S \\ y_{M1} & y_{M2} & y_{M3} & y_{M4} & y_S \\ z_{M1} & z_{M2} & z_{M3} & z_{M4} & z_S \end{bmatrix} \quad \text{(nodal coordinates)}$$

$$[\mathrm{u}1] = \begin{bmatrix} u_{M1} & u_{M2} & u_{M3} & u_{M4} & u_S \\ v_{M1} & v_{M2} & v_{M3} & v_{M4} & v_S \\ w_{M1} & w_{M2} & w_{M3} & w_{M4} & w_S \end{bmatrix} \quad \text{(nodal displacements)}$$

Hints for implementation

1. Using the provided Example 20.2.1, implement the CPP procedure and carefully check the correct values of ξ^1 and ξ^2.
2. Implement the main part of the tangent matrix only together with the residual. Contact must work already!
3. Finish the implementation with the rotational part of the tangent matrix.
4. Entries of $[s]$ have to be symmetric as $[s]^T = [s]$ is valid.
5. Residual set as $[s][ul]$ leads also to a converging computation, but with more iterations.

20.2 Examples

The examples for the 3D NTS discretization start with simple two block setups enabling the check of essential procedures. The complexity of the examples is increased with each example so that the last one deals with bending in two directions.

Since, the focus is on the implementation of the contact element, the nonlinear volume element (elmt7.f) is provided within the finite element code. The structural element has linear shape functions and, therefore, consist of eight nodes.

20.2.1 Two Blocks – 3D Case

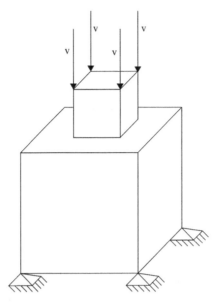

The setup shown here of two block elements is used. A vertical displacement of v is applied in increments (in). Contact is checked between both block elements by a 3D NTS approach. Define the required contact elements (elmt107.f) to solve the contact problem.

Parameters: $E = 2500$, $\nu = 0.0$, $v = -0.2$, $in = 20$, $\varepsilon_N = 25{,}000$

20.2.2 Sliding on a Ramp

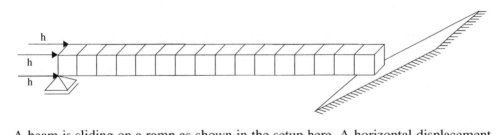

A beam is sliding on a ramp as shown in the setup here. A horizontal displacement of h is applied in increments (in). Contact is checked between beam and ramp by a 3D NTS approach. Define the required contact elements (elmt107.f) to solve the problem.

Parameters: $E = 2500$, $\nu = 0.0$, $h = 5$, $in = 20$, $\varepsilon_N = 25{,}000$

20.2.3 Bending Over a Rigid Cylinder

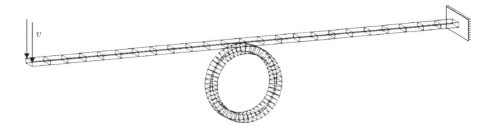

A beam bends over a rigid cylinder as shown in the setup here. A vertical displacement v is applied in increments (in). Contact is checked between beam and cylinder by a 3D NTS discretization.

Parameters: $E = 10^4$, $\nu = 0.0$, $v = 9$, $in = 900$, $\varepsilon_N = 10^3$

20.2.4 Bending Over a Rigid Sphere

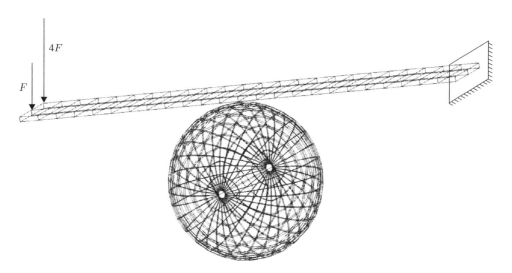

A beam bends over a rigid sphere as shown in the setup here. A vertical force F is applied in increments (in). Contact is checked between beam and sphere by a 3D NTS discretization.

Parameters: $E = 10^4$, $\nu = 0.0$, $F = 0.1$, $in = 250$, $\varepsilon_N = 10^5$

21

Lesson 11
Frictional Node-To-Node (NTN) – `elmt108.f`

The following three sections will deal with frictional contact elements using the already presented Node-To-Node (NTN) and Node-To-Segment (NTS) discretizations with the penalty type of contact enforcement. Here, the setup of the earlier programmed non-frictional elements has to be extended regarding the tangential contact interaction of the tangent matrix and residual vector. The resulting full tangent matrix and residual vector, therefore, consists of a part for the normal interaction and of a part for the tangential interaction. Regarding the tangential part, a return-mapping scheme has to be applied. In case of the NTN discretization the required trial tangential traction becomes rather simple due to being based on the current displacement. The amount of material parameters for `elmt108.f` increases compared to Chapter 13 by the tangential penalty parameter ε_T, the tangential vector τ and the sticking/sliding coefficient μ_s/μ_d, see Figure 21.1.

Similar to the NTN non-frictional contact element, the NTN approach is still applicable to frictional contact problems within the restrictions of the theory given in Chapter 9 in Part I.

21.1 Implementation

Setup of tangent matrix and residual (isw = 3)

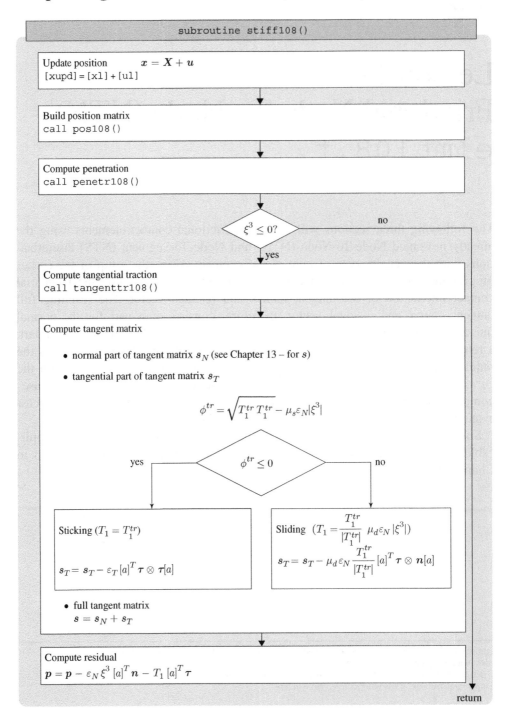

Lesson 11 Frictional Node-To-Node (NTN) – elmt108.f

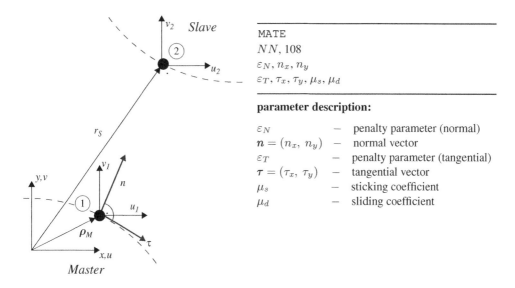

Figure 21.1 Geometry and parameters of the frictional NTN contact element

Remark 21.1.1 *Since the tangential part of the tangent matrix does not have to be symmetric, the global solver used within the Newton scheme (which is usually used for symmetric matrices of conservative problems) should be changed for unsymmetrical matrices.*

Description of subroutines

- subroutine pos108(a)
  Position matrix for $(r_S - \rho_M) = [a]\,[xupd]$
  $$[a] = \begin{bmatrix} -1 & 0 & 1 & 0 \\ 0 & -1 & 0 & 1 \end{bmatrix}$$

- subroutine penetr108(xupd,n,a,penetr)
  Penetration
  $$\xi^3 = [xupd]^T\,[a]^T\,n$$

- subroutine tangenttr108(ul,d,a,$\Delta\xi^1$, T_1^{tr})
  Tangential traction
  $$\Delta\xi^1 = [ul]^T\,[a]^T\,\tau$$
  $$T_1^{tr} = -\varepsilon_T\,\Delta\xi^1$$

Global *FEAP*-arrays

$$[\mathtt{d}] = \begin{bmatrix} \varepsilon_N & n_x & n_y & \varepsilon_T & \tau_x & \tau_y & \mu_s & \mu_d \end{bmatrix} \qquad \text{(material parameters)}$$

$$[\mathtt{xl}] = \begin{bmatrix} x_M & x_S \\ y_M & y_S \end{bmatrix} \qquad \text{(nodal coordinates)}$$

$$[\mathtt{ul}] = \begin{bmatrix} u_M & u_S \\ v_M & v_S \end{bmatrix} \qquad \text{(nodal displacements)}$$

Hints for implementation

1. Enrich the NTN element programmed earlier, see Chapter 13, with the sticking part of the tangent matrix only. Make sure, using provided Examples 21.2.1 and 21.2.2, that full sticking is working.
2. Entries of $[s]$ for sticking have to be symmetric as $[s]^T = [s]$ is valid.
3. Implement the return-mapping algorithm (with corresponding tangent matrices) and carefully check the correct values for stick-slide transition using the frictional patch test in Example 21.2.2.

21.2 Examples

Due to the restrictions of the NTN discretization, especially for frictional problems, the amount of examples using elmt108.f are limited. But, nevertheless, the following two examples will enable us to verify the implementation. First, the same example as in Section 13.2.3 will be discussed extending it to the frictional case, then the frictional patch test follows in order to check the return-mapping scheme more precisely by changing the coefficients of sticking/sliding.

21.2.1 Two Blocks – Frictional Case

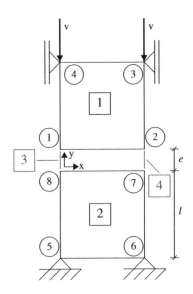

Lesson 11 Frictional Node-To-Node (NTN) – `elmt108.f`

For verification purposes of the programmed routines, the setup shown here of two square plane elements is used. A vertical displacement of v is applied in increments (in). Contact is checked between both planes by a frictional NTN approach. Define the required contact elements (`elmt108.f`) to solve the contact problem. The goal is to study the influence of the sticking/sliding coefficient while varying the Poisson's ratio of the lower block.

Parameters: $E = 10^5$, $\nu = 0.3$, $l = 1$, $v = -0.2$, $in = 20$, $e = 0.01$, $\varepsilon_N = 10^7$, $\boldsymbol{n} = \begin{pmatrix} 0 \\ 1 \end{pmatrix}$, $\varepsilon_T = 10^7$, $\boldsymbol{\tau} = \begin{pmatrix} 1 \\ 0 \end{pmatrix}$, $\mu_s = \mu_d = 0.5$

21.2.2 Frictional Contact Patch Test

The frictional contact patch test is served to check the correct implementation of the return-mapping algorithm:

1. The initially homogenous vertical stress should be correctly split into normal and tangential stresses on the inclined surface.
2. The sticking (no motion) and sliding (motion of contact surfaces) condition should be kinematically verified depending on the inclined angle α and the coefficient of friction ($tg\,\alpha = \mu$ as a threshold value).

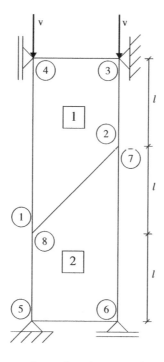

The simplest configuration consists of only two plane elements contacting at an inclined surface of a $45°$ degree angle. Full sticking or sliding is observed, depending on the value of the frictional coefficients. A vertical displacement of v is applied in increments (in). Contact is checked between both planes by a frictional NTN approach.

Goals:

1. Define the required contact elements (elmt108.f) to solve the contact problem.
2. Verify the contact behavior for the frictional coefficients being below the value 1.0 (sliding should be detected).
3. Verify the behavior for the frictional coefficients being above 1.0 (full sticking should be detected).

Parameters: $E = 10^5$, $\nu = 0.3$, $l = 1$, $v = -0.2$, $in = 20$, $\varepsilon_N = 10^8$, $\mathbf{n} = \begin{pmatrix} -1 \\ 1 \end{pmatrix}$, $\varepsilon_T = 10^8$, $\boldsymbol{\tau} = \begin{pmatrix} 1 \\ 1 \end{pmatrix}$

22

Lesson 12
Frictional Node-To-Segment (NTS) – `elmt109.f`

The chapter deals with a more general contact formulation for large sliding problems – a frictional NTS discretization, see Figure 22.1. Compared with the NTN approach, geometrical parameters such as normal and tangent vectors, the amount of sliding (measure of tangential interaction) is computed within the contact algorithm for the frictional NTS contact element. Compared to the frictional NTN, the frictional NTS `elmt109.f` enables larger tangential displacements, thus, a better approximation has to be provided regarding the trial tangential traction T_1^{tr} which is based on a simplified backward Euler scheme with no update for the sticking point once initialized. The initialization runs, hereby, with history variables set up on global level and associated with the load steps (0) and (n) for `nh1` and `nh2`, respectively. After each load step, the entries at pointer `nh2` are written on entries at pointer `nh1`. Moreover, the frictional NTS possesses additional rotational entries for the tangential part of the tangent matrix.

22.1 Implementation

Setup of tangent matrix and residual (`isw = 3`)

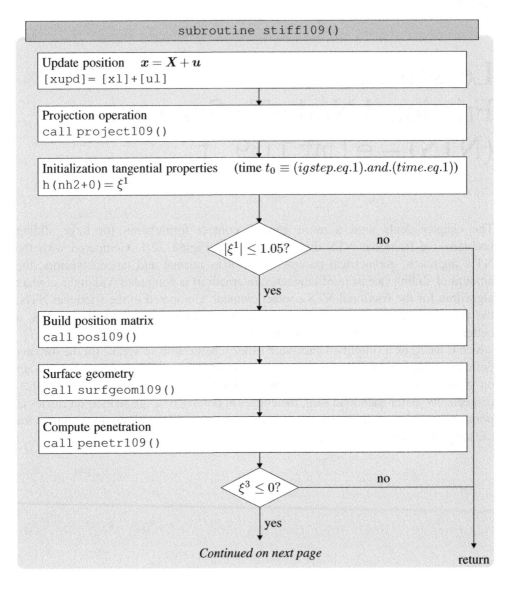

| Continued: stiff109() |

From previous page
↓

Compute convective coordinate increment
call tangentvel109()
↓

Compute tangential traction
call tangenttr109()
↓

Compute tangent matrix
- normal part of tangent matrix s_N (see Chapter 16 – for s)
- tangential part of tangent matrix s_T

$$\phi^{tr} = \sqrt{T_1^{tr}\, T_1^{tr}\, m^{11}} - \mu_s \varepsilon_N |\xi^3|$$

$\phi^{tr} \leq 0$?

yes →

Sticking $(T_1 = T_1^{tr})$

$$s_{T_1} = \varepsilon_T [a]^T \boldsymbol{\tau} \otimes \boldsymbol{\tau}[a]$$

$$s_{T_2} = T_1 m^{11} [da]^T \boldsymbol{\tau} \otimes \boldsymbol{\tau}[a]$$

$$s_T = -s_{T_1} + s_{T_2} + s_{T_2}^T$$

no →

Sliding $(T_1 = \dfrac{T_1^{tr}}{|T_1^{tr}|} \mu_d \varepsilon_N |\xi^3| \sqrt{m_{11}})$

$$s_{T_1} = \mu_d \varepsilon_N \frac{T_1^{tr}}{|T_1^{tr}|} [a]^T \boldsymbol{\tau} \otimes \boldsymbol{n}[a]$$

$$s_{T_2} = \mu_d \varepsilon_N |\xi^3| \frac{T_1^{tr}}{|T_1^{tr}|} \sqrt{m^{11}} [da]^T \boldsymbol{\tau} \otimes \boldsymbol{\tau}[a]$$

$$s_T = -s_{T_1} + s_{T_2} + s_{T_2}^T$$

- full tangent matrix
$s = s_N + s_T$

↓

Compute residual
$$\boldsymbol{p} = \boldsymbol{p} - \varepsilon_N\, \xi^3\, [a]^T\, \boldsymbol{n} - T_1 \sqrt{m^{11}} [a]^T \boldsymbol{\tau}$$

return

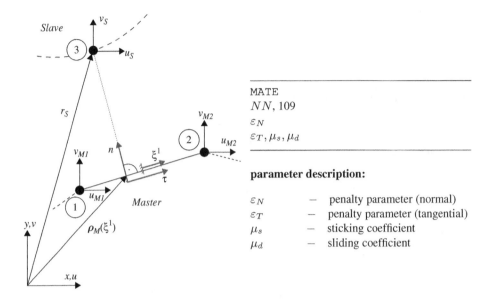

Figure 22.1 Geometry and parameters of the frictional NTS contact element

Description of subroutines

- subroutine project109(xupd, ξ^1)
  Closest point projection of $\boldsymbol{r}_S$ on $\boldsymbol{\rho}_M$

$$F(\xi^1) = \|\boldsymbol{r}_S - \boldsymbol{\rho}_M(\xi^1)\| \to min$$

for linear NTS in 2D (see Section 7.4.1 in Part I):

$$\xi^1 = \frac{2[x_S(x_{M2} - x_{M1}) + y_S(y_{M2} - y_{M1})] - x_{M2}^2 - y_{M2}^2 + x_{M1}^2 + y_{M1}^2}{(x_{M2} - x_{M1})^2 + (y_{M2} - y_{M1})^2}$$

- subroutine pos109(a, da, ξ^1)
  Position matrix for $(\boldsymbol{r}_S - \boldsymbol{\rho}_M) = [a][xupd]$ and derivative
  call shape109()

$$[a] = \begin{bmatrix} -N_1 & 0 & -N_2 & 0 & 1 & 0 \\ 0 & -N_1 & 0 & -N_2 & 0 & 1 \end{bmatrix}$$

$$[da] = \frac{\partial[a]}{\partial \xi^1} = \begin{bmatrix} -N_{1,\xi^1} & 0 & -N_{2,\xi^1} & 0 & 0 & 0 \\ 0 & -N_{1,\xi^1} & 0 & -N_{2,\xi^1} & 0 & 0 \end{bmatrix}$$

- subroutine shape109(ξ^1, $\boldsymbol{N}(2,2)$)
  Shape functions and derivatives

$$[shape] = \begin{bmatrix} N_1 & N_2 \\ N_{1,\xi^1} & N_{2,\xi^1} \end{bmatrix}, \text{ with } N_1 = \frac{1}{2}(1 - \xi^1), N_2 = \frac{1}{2}(1 + \xi^1)$$

Lesson 12 Frictional Node-To-Segment (NTS) – elmt109.f

- subroutine surfgeom109(xupd,da,$\boldsymbol{n}$,$\boldsymbol{\tau}$,$\boldsymbol{\rho}_{,\xi^1}$,$m_{11}$,$m^{11}$,detm)
  Surface vector
$$\boldsymbol{\rho}_{,\xi^1} = -[da][xupd]$$

Metric tensor
$$m_{11} = \boldsymbol{\rho}_{,\xi^1} \cdot \boldsymbol{\rho}_{,\xi^1}$$

$$m^{11} = \frac{1}{m_{11}}$$

$$det\boldsymbol{m} = m_{11}$$

Normal vector
$$\boldsymbol{n} = -\frac{1}{2\sqrt{det\boldsymbol{m}}}\begin{pmatrix} y_{M2} - y_{M1} \\ x_{M1} - x_{M2} \end{pmatrix}$$

Tangent vector
$$\boldsymbol{\tau} = \frac{\boldsymbol{\rho}_{,\xi^1}}{\sqrt{det\boldsymbol{m}}}$$

- subroutine penetr109(xupd,$\boldsymbol{n}$,a,penetr)
  Penetration
$$\xi^3 = [xupd]^T[a]^T\boldsymbol{n}$$

- subroutine tangentvel109(ξ^1,$\Delta\xi^1$)
  Velocity with finite difference scheme ($\Delta\xi^1 = \xi^1_{(n+1)} - \xi^1_{(n)}$)
$$\Delta\xi^1 = \xi^1_{(n+1)} - h(nh1+0)$$

ξ^1 for the next iteration
$$h(nh2+0) = \xi^1_{n+1}$$

- subroutine tangenttr109(d,$\Delta\xi^1$,m_{11},T_1^{tr})
  Tangential traction
$$T_1^{tr} = -\varepsilon_T m_{11} \Delta\xi^1$$

Global *FEAP*-arrays

$[\mathtt{d}] = \begin{bmatrix} \varepsilon_N & \varepsilon_T & \mu_s & \mu_d \end{bmatrix}$ (material parameters)

$[\mathtt{xl}] = \begin{bmatrix} x_{M1} & x_{M2} & x_S \\ y_{M1} & y_{M2} & y_S \end{bmatrix}$ (nodal coordinates)

$[\mathtt{ul}] = \begin{bmatrix} u_{M1} & u_{M2} & u_S \\ v_{M1} & v_{M2} & v_S \end{bmatrix}$ (nodal displacements)

Hints for implementation

1. Enrich the NTS non-frictional element programmed earlier, see Chapter 16 with the sticking part of the tangent matrix only. Make sure when using the frictional contact patch test in Example 22.2.2 that the sticking part is working correctly.
2. Entries of $[s]$ for the sticking case have to symmetric as $[s]^T = [s]$ is valid.
3. Using the parameters for sliding of $\mu_s = \mu_d = 0.0$, carefully check the amount of sliding on the master surface $\Delta \xi^1$. In this step the correct transfer of history variables is checked.
4. Implement fully the return-mapping algorithm using the frictional contact patch test from Example 22.2.2. Verify the correct threshold value of $\mu_s = \mu_d$ for the start of sticking-sliding transition.

22.2 Examples

The following examples are intended to demonstrate the capabilities of elmt109.f with large tangential displacements, even including sliding.

22.2.1 Two Blocks

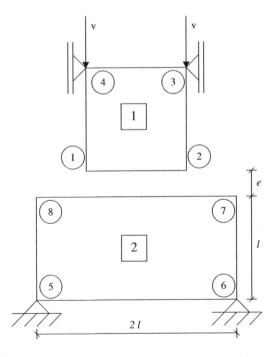

For verification purposes of the programmed routines, the setup shown here of two plane elements is used. A vertical displacement of v is applied in increments (in). Contact is checked between both planes by a frictional NTS approach. Define the

Lesson 12 Frictional Node-To-Segment (NTS) – elmt109.f

required contact elements (elmt109.f) to solve the problem. The goal is to study the influence of the sticking/sliding coefficient while varying the Poisson's ratio of the lower block.

Parameters: $E = 10^5, \nu = 0.3, l = 1, v = -0.2, in = 20, e = 0.01, \varepsilon_N = 10^7, \varepsilon_T = 10^7, \mu_s = 0.5, \mu_d = 0.5$

22.2.2 Frictional Contact Patch Test

The frictional contact patch test is an *obligatory verification test* to check the correct implementation of the return-mapping algorithm working together with the Closest Point Projection procedure, see also the Example in 21.2.2.

1. The initially homogenous vertical stress should be correctly split into normal and tangential stresses on the inclined surface.
2. The sticking (no motion) and sliding (motion of contact surfaces) condition should be kinematically verified depending on the inclined angle α and the coefficient of friction ($tg\,\alpha = \mu$ as a threshold value).

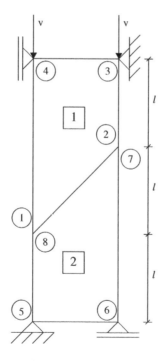

The simplest configuration consists of only two plane elements contacting at an inclined surface of a 45° angle. Full sticking or sliding is observed, depending on the value of the frictional coefficients. A vertical displacement of v is applied in increments (in). Contact is checked between both planes by a frictional NTS approach.

Goals:

1. Define the required contact elements (elmt109.f) to solve the contact problem.
2. Verify the contact behavior for the frictional coefficients that are below the value 1.0 (sliding should be detected).
3. Verify the behavior for the frictional coefficients that are above 1.0 (full sticking should be detected).

Parameters: $E = 10^5$, $\nu = 0.0...2.0$, $l = 1$, $v = -0.2$, $in = 20$, $\varepsilon_N = \varepsilon_T = 10^8$

22.2.3 Block and Inclined Rigid Surface – Different Boundary Condition

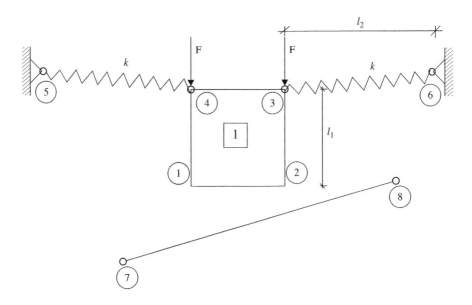

Next, on the setup shown here a square plane attached to springs over an inclined plane is used. A vertical force of F is applied in increments (in). Contact is now checked between both planes by a frictional NTS approach.

Vary μ_d and μ_s from 0.0 to 0.8 by increments of 0.1. What can be seen? Study the start of sticking-sliding transition carefully.

Parameters: $E = 10^7$, $\nu = 0.3$, $k = 8 \cdot 10^4$, $l_1 = 1$, $l_2 = 1$, $F = -10,000$, $in = 200$, $\varepsilon_N = 10^8$, $\varepsilon_T = 10^8$

22.2.4 Generalized 2D Euler–Eytelwein Problem

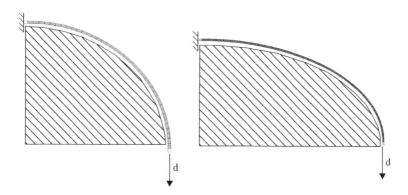

The last example is the generalized 2D Euler–Eytelwein problem where plane elements are used to model a rope-like structure contacting a rigid circular or elliptic surface. A curved rigid boundary is represented by fixed master segments. A vertical displacement of d is applied in increments (in). Contact is again checked between the surface and the rope by a frictional NTS approach. Vary $\mu_d = \mu_s = \mu$ from 0.0 to 0.5 by increments of 0.05.

Goals:

1. Modify the CPP to $\xi^1 \leq 1.00$ and the penetration check to a permanent contact!
2. Consider the reaction forces at both ends of the rope. Which one has to be larger?
3. Compare the resulting Euler–Eytelwein relation $\frac{T}{T_0}$ for the frictional coefficient range of $\mu \in (0.0; 0.5)$ with the analytically given result ($\frac{T}{T_0} = e^{\mu \cdot \frac{\pi}{2}}$) in Section 8.4 in Part I. Perform a diagram plot in this case.

Parameters: $E = 21,000$, $\nu = 0.0$, $d = -0.001$, $in = 1$, $\varepsilon_N = 10^4$, $\varepsilon_T = 10^4$

Results of computation are given in Fig. 22.2: One should understand that the analysis presented in Section 8.4 is obtained for a rope that is not capable of carrying the bending moment ("moment-free"). For the current verification, however, we are just using a plane finite element to model a rope. So it is surprising that in the case of a circle the verification is quite good for such a rough model. In the case of an ellipse, the bending stresses are obviously preventing being close to the analytical solution. An advanced rope model, together with an advanced curve-to-surface contact algorithm, are necessary in this case to obtain a good correlation.

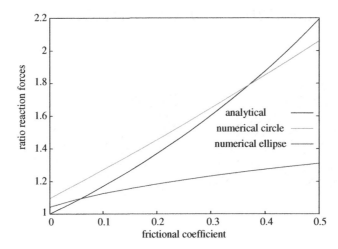

Figure 22.2 Ratio between reaction forces of both rope ends for different frictional coefficients

23

Lesson 13
Frictional Higher Order NTS – elmt110.f

The last element for the implementation of frictional contact is the frictional higher order NTS element that enables, as presented earlier for frictionless contact in Chapter 19, to use the Lagrange, Lobatto and Bezier classes of quadratic shape functions. In this case the CPP procedure no longer has a closed form solution and it is necessary to solve the CPP problem inside the element subroutine locally via Newton's method. In addition, all parts of the tangent matrix, including curvature parts, should be implemented compared to the linear NTS, see Chapter 16.

The trial tangential traction T_1^{tr} is based on a more advanced update scheme with the sticking point and tangential traction being updated for each load step. This update scheme in particular is required for elmt110.f, since, it enables curved surfaces, see Figure 23.1.

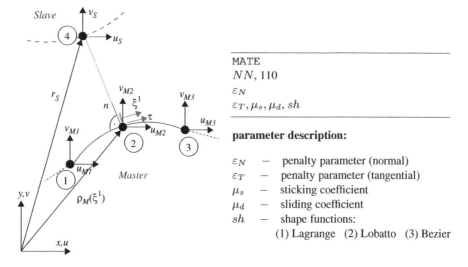

Figure 23.1 Geometry and parameters of the frictional higher order NTS contact element

Introduction to Computational Contact Mechanics: A Geometrical Approach, First Edition.
Alexander Konyukhov and Ridvan Izi.
© 2015 John Wiley & Sons, Ltd. Published 2015 by John Wiley & Sons, Ltd.
Companion Website: www.wiley.com/go/Konyukhov

23.1 Implementation

Setup of tangent matrix and residual (isw = 3)

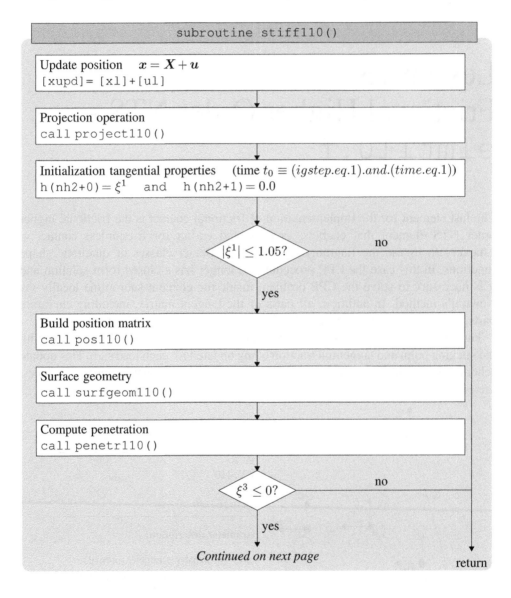

Lesson 13 Frictional Higher Order NTS – `elmt110.f`

Continued: `stiff110()`

From previous page

Compute convective coordinate increment
`call tangentvel110()`

Compute tangential traction
`call tangenttr110()`

Compute tangent matrix

- normal part of tangent matrix
  $s_N = s_{Nm} + s_{Nr} + s_{Nc}$ (see Chapters 16 – for $s_{Nm} + s_{Nr}$ and 19 – for s_{Nc})
- tangential part of tangent matrix s_T

$$\phi^{tr} = \sqrt{T_1^{tr} \, T_1^{tr} \, m^{11}} - \mu_s \varepsilon_N |\xi^3|$$

Decision: $\phi^{tr} \leq 0$?

yes → Sticking ($T_1 = T_1^{tr}$)

$$s_{T_1} = \varepsilon_T \, [a]^T \boldsymbol{\tau} \otimes \boldsymbol{\tau}[a]$$

$$s_{T_2} = T_1 \, m^{11} \, [a_{,\xi^1}]^T \boldsymbol{\tau} \otimes \boldsymbol{\tau}[a]$$

$$s_{T_3} = T_1 \, h^{11} \, \sqrt{m_{11}} \, [a]^T (\boldsymbol{\tau} \otimes \boldsymbol{n} + \boldsymbol{n} \otimes \boldsymbol{\tau})[a]$$

$$s_T = -s_{T_1} + s_{T_2} + s_{T_2}^T + s_{T_3}^T$$

no → Sliding $\left(T_1 = \dfrac{T_1^{tr}}{|T_1^{tr}|} \mu_d \varepsilon_N |\xi^3| \sqrt{m_{11}}\right)$

$$s_{T_1} = \mu_d \varepsilon_N \frac{T_1^{tr}}{|T_1^{tr}|} [a]^T \boldsymbol{\tau} \otimes \boldsymbol{n}[a]$$

$$s_{T_2} = \mu_d \varepsilon_N |\xi^3| \frac{T_1^{tr}}{|T_1^{tr}|} \sqrt{m^{11}} \, [a_{,\xi^1}]^T \boldsymbol{\tau} \otimes \boldsymbol{\tau}[a]$$

$$s_{T_3} = \mu_d \varepsilon_N |\xi^3| \frac{T_1^{tr}}{|T_1^{tr}|} h^{11} m_{11} [a]^T (\boldsymbol{\tau} \otimes \boldsymbol{n} + \boldsymbol{n} \otimes \boldsymbol{\tau})[a]$$

$$s_T = -s_{T_1} + s_{T_2} + s_{T_2}^T + s_{T_3}^T$$

- full tangent matrix
  $s = s_N + s_T$

Compute residual
$$p = p - \varepsilon_N \, \xi^3 \, [a]^T \, \boldsymbol{n} - T_1 \sqrt{m^{11}} [a]^T \boldsymbol{\tau}$$

return

Description of subroutines

- subroutine project110(xupd,ξ^1,d)
  Closest point projection of r_S on ρ_M

$$F(\xi^1) = \|r_S - \rho_M(\xi^1)\| \to min$$

Newton's method with $\xi^1_{(n+1)} = \xi^1_{(n)} + \Delta\xi^1_{(n)}$ and $\xi^1_{(0)} = 0$
where

$$\Delta\xi^1_{(n)} = -(F'')^{-1}_{(n)} F'_{(n)} \quad \text{with} \quad F(\xi^1) = \frac{1}{2}(r_S - \rho_M)\cdot(r_S - \rho_M)$$

$$F' = (r_S - \rho_M(\xi^1))\cdot\frac{\partial\rho_M}{\partial\xi^1}$$

$$F'' = (r_S - \rho_M(\xi^1))\cdot\frac{\partial^2\rho_M}{\partial^2\xi^1} - \frac{\partial\rho_M}{\partial\xi^1}\cdot\frac{\partial\rho_M}{\partial\xi^1}$$

$$\Delta\xi^1_{(n)} = -\frac{(r_S - \rho_M)\cdot\rho_{,\xi^1}}{(r_S - \rho_M)\cdot\rho_{,\xi^1\xi^1} - \rho_{,\xi^1}\cdot\rho_{,\xi^1}}$$

  call pos110()
  call surfgeom110()

$$\Delta\xi^1_{(n)} = -\frac{f}{e - m_{11}} \quad \text{with} \quad f = \rho_{,\xi^1}[a][xupd], \ e = \rho_{,\xi^1\xi^1}[a][xupd]$$

- subroutine pos110(a,$a_{,\xi^1}$,$a_{,\xi^1\xi^1}$,ξ^1,d)
  Position matrix for $(r_S - \rho_M) = [a][xupd]$ and derivatives
  call shape110()

$$[a] = \begin{bmatrix} -N_1 & 0 & -N_2 & 0 & -N_3 & 0 & 1 & 0 \\ 0 & -N_1 & 0 & -N_2 & 0 & -N_3 & 0 & 1 \end{bmatrix}$$

$$[a_{,\xi^1}] = \frac{\partial[a]}{\partial\xi^1} = \begin{bmatrix} -N_{1,\xi^1} & 0 & -N_{2,\xi^1} & 0 & -N_{3,\xi^1} & 0 & 0 & 0 \\ 0 & -N_{1,\xi^1} & 0 & -N_{2,\xi^1} & 0 & -N_{3,\xi^1} & 0 & 0 \end{bmatrix}$$

$$[a_{,\xi^1\xi^1}] = \frac{\partial^2[a]}{\partial\xi^{1^2}} = \begin{bmatrix} -N_{1,\xi^1\xi^1} & 0 & -N_{2,\xi^1\xi^1} & 0 & -N_{3,\xi^1\xi^1} & 0 & 0 & 0 \\ 0 & -N_{1,\xi^1\xi^1} & 0 & -N_{2,\xi^1\xi^1} & 0 & -N_{3,\xi^1\xi^1} & 0 & 0 \end{bmatrix}$$

- subroutine shape110(ξ^1,$N(3,3)$,d)
  Shape functions and derivatives

$$[shape] = \begin{bmatrix} N_1 & N_2 & N_3 \\ N_{1,\xi^1} & N_{2,\xi^1} & N_{3,\xi^1} \\ N_{1,\xi^1\xi^1} & N_{2,\xi^1\xi^1} & N_{3,\xi^1\xi^1} \end{bmatrix},$$

Lesson 13 Frictional Higher Order NTS – elmt110.f

with

1. Lagrange: $N_1 = \frac{1}{2}\xi^1(\xi^1 - 1)$, $N_2 = 1 - {\xi^1}^2$, $N_3 = \frac{1}{2}\xi^1(\xi^1 + 1)$

2. Lobatto: $N_1 = \frac{1}{2}(1 - \xi^1)$, $N_2 = \frac{1}{2}\sqrt{\frac{3}{2}}((\xi^1)^2 - 1)$, $N_3 = \frac{1}{2}(1 + \xi^1)$

3. Bezier: $N_1 = \frac{1}{4}(1 - \xi^1)^2$, $N_2 = \frac{1}{2}(1 - \xi^1)(1 + \xi^1)$, $N_3 = \frac{1}{4}(1 + \xi^1)^2$

- subroutine surfgeom110(xupd, $a_{,\xi^1}$, $a_{,\xi^1\xi^1}$, $\boldsymbol{n}$, $\boldsymbol{\tau}$, $\boldsymbol{\rho}_{,\xi^1}$, $\boldsymbol{\rho}_{,\xi^1\xi^1}$, m^{11}, m_{11}, h^{11}, $det\boldsymbol{m}$)

Surface vector

$$\boldsymbol{\rho}_{,\xi^1} = -[a_{,\xi^1}][xupd]$$

Metric tensor

$$m_{11} = \boldsymbol{\rho}_{,\xi^1} \cdot \boldsymbol{\rho}_{,\xi^1}$$

$$m^{11} = \frac{1}{m_{11}}$$

$$det\boldsymbol{m} = m_{11}$$

Normal vector

$$\boldsymbol{n} = \frac{1}{\sqrt{m_{11}}}\begin{pmatrix} -\rho_{,\xi^1}(2) \\ \rho_{,\xi^1}(1) \end{pmatrix}$$

Second derivative of $\rho(\xi^1)$

$$\boldsymbol{\rho}_{,\xi^1\xi^1} = -[a_{,\xi^1\xi^1}][xupd]$$

Tangent vector

$$\boldsymbol{\tau} = \frac{\boldsymbol{\rho}_{,\xi^1}}{\sqrt{m_{11}}}$$

Curvature tensor

$$h^{11} = \frac{\boldsymbol{\rho}_{,\xi^1\xi^1} \cdot \boldsymbol{n}}{(\boldsymbol{\rho}_{,\xi^1} \cdot \boldsymbol{\rho}_{,\xi^1})^2}$$

- subroutine penetr110(xupd, $\boldsymbol{n}$, a, penetr)

Penetration

$$\xi^3 = [xupd]^T[a]^T\boldsymbol{n}$$

- subroutine tangentvel110(ξ^1, $\Delta\xi^1$)

Velocity with finite difference scheme ($\Delta\xi^1 = \xi^1_{(n+1)} - \xi^1_{(n)}$)

$$\Delta\xi^1 = \xi^1_{(n+1)} - h(nh1 + 0)$$

ξ^1 for the next iteration

$$h(nh2 + 0) = \xi^1_{(n+1)}$$

- subroutine tangenttr110 ($\texttt{d}, \Delta\xi^1, m_{11}, T_1^{tr}$)
  Tangential traction with finite difference scheme

$$T_{1(n+1)}^{tr} = h(nh1+1) - \varepsilon_T m_{11} \Delta\xi^1$$

T_1^{tr} for the next iteration

$$h(nh2+1) = T_{1(n+1)}^{tr}$$

Global *FEAP*-arrays

$$[\texttt{d}] = \begin{bmatrix} \varepsilon_N & \varepsilon_T & \mu_s & \mu_d & sh \end{bmatrix} \qquad \text{(material parameters)}$$

$$[\texttt{xl}] = \begin{bmatrix} x_{M1} & x_{M2} & x_{M3} & x_S \\ y_{M1} & y_{M2} & y_{M3} & y_S \end{bmatrix} \qquad \text{(nodal coordinates)}$$

$$[\texttt{ul}] = \begin{bmatrix} u_{M1} & u_{M2} & u_{M3} & u_S \\ v_{M1} & v_{M2} & v_{M3} & v_S \end{bmatrix} \qquad \text{(nodal displacements)}$$

Hints for implementation

1. Enrich already programmed earlier NTS frictional element, see Chapter 22, with higher order shape functions.
2. Using the provided Example 23.2.1 carefully check the CPP procedure for correct values of ξ^1.
3. Finish the implementation with the rotational and curvature part of the tangent matrix.
4. Entries of $[s]$ for the sticking case have to symmetric as $[s]^T = [s]$ is valid.
5. Using parameters for sliding of $\mu_s = \mu_d = 0.0$, carefully check the amount of sliding on the master surface $\Delta\xi^1$. In this step the correct transfer of history variables is checked.

23.2 Examples

The frictional higher order NTS element is tested for the same setups as before for the linear case, see Chapter 22. But, as for dealing with higher order elements before in Chapter 19 the examples are conducted with curved surfaces where a comparable geometry is intended for the different shape function classes.

23.2.1 Two Blocks

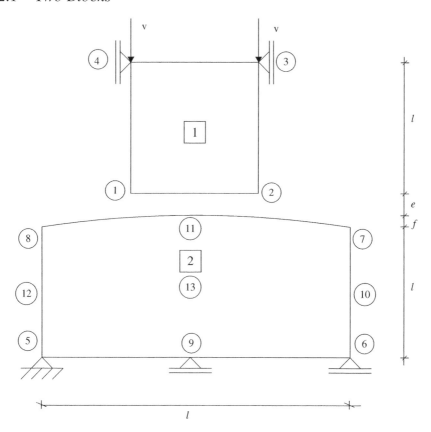

The setup shown here of two plane elements is used. A vertical displacement of v is applied in increments (in). Contact is checked between both contacting segments by a frictional higher order NTS approach. Define the required contact elements (elmt110.f) to solve the problem. The goal is to study the influence of the sticking/sliding coefficient while varying the Poisson's ratio of the lower block. Plot the initial and final configuration for each shape function class. Compared to Example 22.2.1, a horizontal displacement of node 7 and 8 has to be observed even for $\nu = 0.0$!

Parameters: $E = 10^5$, $\nu = 0.3$, $l = 1$, $v = -0.2$, $in = 20$, $e = 0.015$, $f = 0.2$, $\varepsilon_N = 10^6$, $\varepsilon_T = 10^6$, $\mu_s = 1.0$, $\mu_d = 1.0$

Results of computation for the Lagrange, Lobatto and Bezier shape functions are given in Fig. 23.2:

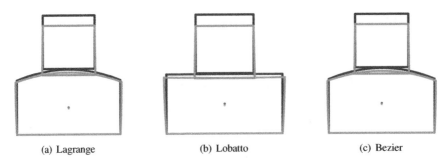

(a) Lagrange (b) Lobatto (c) Bezier

Figure 23.2 Initial and final configuration for sticking with (a) Lagrange, (b) Lobatto and (c) Bezier classes of shape functions, respectively

23.2.2 Block and Inclined Rigid Surface – Different Boundary Condition

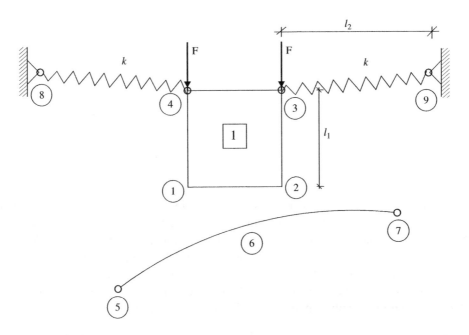

Next, the setup shown here of a plane attached to springs over an inclined segment is used. A vertical force of F is applied in increments (in). Contact is now checked between both contacting segments by a frictional higher order NTS discretization.

Vary μ_d and μ_s from 0.0 to 0.8 by increments of 0.1. What can be seen? Plot the initial and final configuration for each shape function class.

Parameters: $E = 10^6$, $\nu = 0.3$, $k = 8 \cdot 10^4$, $l_1 = 1$, $l_2 = 1$, $F = -200,000$, $in = 200$, $\varepsilon_N = 10^8$, $\varepsilon_T = 10^8$

Results of computation for the Lagrange, Lobatto and Bezier shape functions (sticking case) are given in Fig. 23.3:

Lesson 13 Frictional Higher Order NTS – elmt110.f

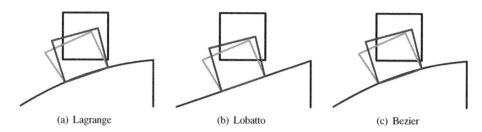

(a) Lagrange   (b) Lobatto   (c) Bezier

Figure 23.3 Initial, intermediate and final configuration for sliding with (a) Lagrange, (b) Lobatto and (c) Bezier classes of shape functions

24

Lesson 14
Transient Contact Problems

All the previous chapters in Part II of this book were devoted to the implementation and verification tasks of contact elements for static, or quasi-static problems. This chapter aims to show the implementation of the transient problem. For static contact problems, the most important parts are correct discretization and implementation of both the tangent matrix and residual being written in the covariant form. In general, for dynamic problems only the correct mass matrix should be additionally implemented. That is why the extension into the dynamic problems is straightforward and all programmed algorithms can be carried out on an element level.

A set of problems, however, is typical for the global level of implementation of transient problems. The choices are between the implicit or the explicit time integration scheme, stability of the time integration scheme, conservation of the energy and momentum, global and local tolerance and so on. These problems are discussed in special texts on modeling of dynamic problems in structural mechanics and are outside the scope of this book. Based on the mentioned criteria, there is a large variety of time integration schemes available. Here, we are going to use the most common implicit scheme – the Newmark time integration scheme given in the flowchart that follows, see Figure 24.1. Parameters of β and γ control the output and are chosen by 0.25 and 0.5, respectively, in order to fulfill the energy conservation property, see details in Hughes (2000). Due to the fact that time integration schemes deal on the global level of a finite element code, the implementation will not be discussed further, but presented next. The full setup of dynamic problems includes a nodal velocity vector $\dot{d}$ and nodal acceleration vector $\ddot{d}$ together with the nodal displacement vector d used before for static and quasi-static problems. Further system characteristics, such as the mass matrix M, have to be taken into account besides the stiffness matrix K already considered earlier for static problems. The main focus is to show the applicability of the programmed contact elements for the transient analysis.

Introduction to Computational Contact Mechanics: A Geometrical Approach, First Edition.
Alexander Konyukhov and Ridvan Izi.
© 2015 John Wiley & Sons, Ltd. Published 2015 by John Wiley & Sons, Ltd.
Companion Website: www.wiley.com/go/Konyukhov

24.1 Implementation

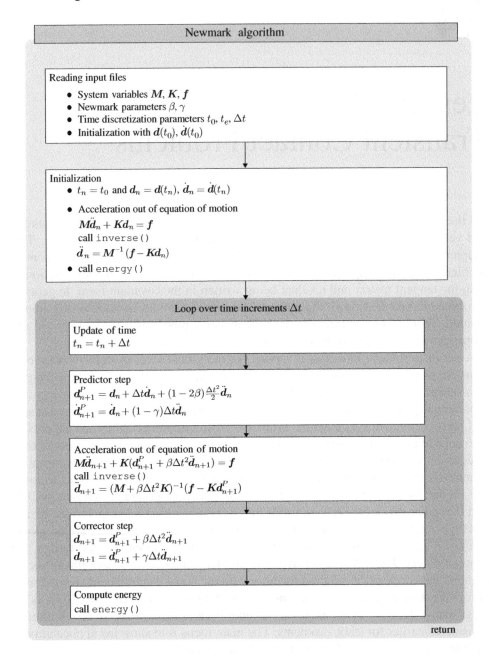

Lesson 14 Transient Contact Problems

Description of subroutines
- subroutine energy(**M**, **K**, $\dot{\mathbf{d}}$, $\dot{\mathbf{d}}$)
  kinetical energy
  $$T = \frac{1}{2}\dot{d}^T M \dot{d}$$
  potential energy
  $$V = \frac{1}{2}d^T K d$$
- subroutine inverse(**M**, **M**$^{-1}$)
  $$M^{-1} = \frac{1}{\det M}\begin{bmatrix} M_{22} & -M_{12} \\ -M_{12} & M_{11} \end{bmatrix}$$

24.2 Examples

As discussed within Example 17.2.2, structural problems in statics require Dirichlet boundary conditions. Therefore, the same example is first discussed in dynamics using the Newmark algorithm for the frictional and non-frictional cases. The main goal is to see the applicability of the discussed elements during postprocessing.

24.2.1 Block and Inclined Rigid Surface – Non-Frictional Case

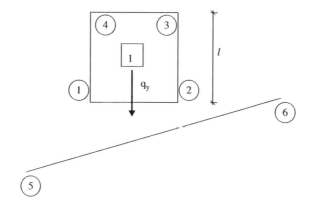

The setup shown here of a square plane over an inclined plane is used. A vertical volume load of q_y is applied. Contact is now checked between both planes by NTS contact elements (elmt103.f).
Goals:

1. Observe the dynamic behaviour of the system.
2. Study the influence of the time step size on the solution.

Parameters: $E = 10^7$, $\nu = 0.3$, $l = 1$, $\rho = 10$, $q_y = -1$, $\varepsilon_N = 10^8$

24.2.2 Block and Inclined Rigid Surface – Frictional Case

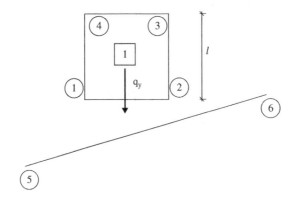

The same setup is used for the frictional example. Contact is now checked between both planes by a frictional NTS discretization (elmt109.f). Vary μ_d and μ_s from 0.0 to 0.8 by increments of 0.1. What can be seen within the postprocessing?

Parameters: $E = 10^7$, $\nu = 0.3$, $l = 1$, $\rho = 10$, $q_y = -1$, $\varepsilon_N = 10^8$, $\varepsilon_T = 10^8$, $\mu_s = 1.0$, $\mu_d = 0.0$

24.2.3 Moving Pendulum with Impact – Center of Percussion

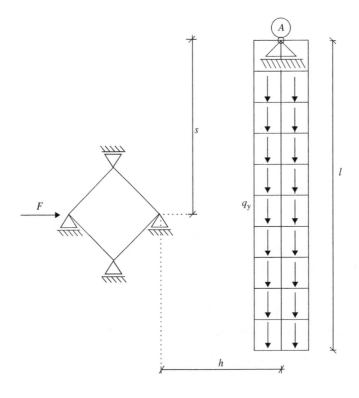

Lesson 14 Transient Contact Problems

The example shown here is a pendulum with a volume load q_y exposed to impact by a moving barrier. The barrier is moved due to a horizontal force of F while consisting of a density of ρ. Its position is given at a distance of h in horizontal and s in vertical direction, respectively.

Goals:

1. Define NTS contact elements (elmt103.f) all over the contacting surface of the pendulum.
2. Change the position of the barrier in vertical direction ($0 \leq s \leq l$) and plot the resulting horizontal displacement of the first non-penetrating time step after the impact for position s. For which position s^* does the horizontal displacement vanish? The goal is to find the center of percussion of the pendulum, see the solution in Section 8.3 in Chapter 8, Part I.
3. Remesh the pendulum so that for s^* a NTN discretization is applicable. Is a movement of the pendulum observed for $s = s^*$?

Parameters: $E = 10^7, \nu = 0.3, l = 30, h = 4, \rho = 10, F = 100, q_y = -1, \varepsilon_N = 10^5$

Results of computation are given in Fig. 24.1:

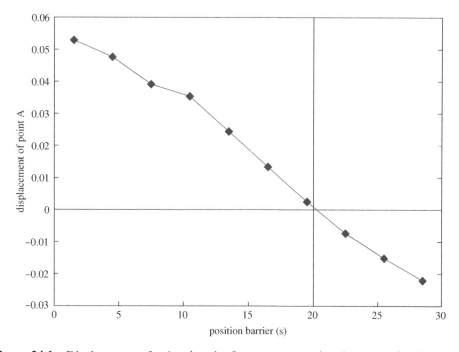

Figure 24.1 Displacement of point A at the first non-penetrating time step after impact vs. the position s of the barrier

The Figure 24.1 illustrating the displacement of the moving support point A vs. position of the impactor s of the barrier. This allows to define the center of percussion as an intersection of the curve with zero line. The result is obtained as approximately 20, which is well enough correlated with the theoretical value as $2/3l = 20$.

Appendix A

Numerical integration

In the case of numerical computation of integrals, different formulas are available as Gauss or Lobatto integration formulas. In all cases this leads to an approximation of the integral with a remaining error R. Depending on the order of the integrated polynomial and the amount of integration points N this error R vanishes and the integral is exactly evaluated. Thus, the statement holds

$$\int_a^b \tilde{g}(x)dx = \int_{-1}^1 g(\xi)d\xi = \sum_{i=1}^N g(\xi_i)w_i + R \approx \sum_{i=1}^N g(\xi_i)w_i \qquad (A.1)$$

where ξ_i represents the abscissa of the quadrature formula and w_i the corresponding weight. In case of the aforementioned Gauss and Lobatto integration formula this may look like

Gauss

N/ip	i	ξ_i	w_i
1	1	0	2
2	1	$-1/\sqrt{3}$	1
	2	$+1/\sqrt{3}$	1
3	1	$-\sqrt{3/5}$	5/9
	2	0	8/9
	3	$+\sqrt{3/5}$	5/9
⋮	⋮	⋮	⋮

N/ip	i	ξ_i	w_i
2	1	-1	1
	2	1	1
3	1	-1	1/3
	2	0	4/3
	3	1	1/3
4	1	-1	1/6
	2	$-\sqrt{\frac{1}{5}}$	5/6
	3	$\sqrt{\frac{1}{5}}$	5/6
	4	1	1/6
$\vdots$	$\vdots$	$\vdots$	$\vdots$

Lobatto

Here, the integration formulas allow us to integrate exactly the polynomials up to the following order $\begin{cases} 2N-1 \\ 2N-3 \end{cases}$ for $\begin{matrix} \text{Gauss} \\ \text{Lobatto} \end{matrix}$ expressed within a table gives

Type	polynomial order	N = 1	N = 2	N = 3
Gauss	$2N-1$	1	3	5
Lobatto	$2N-3$	-	1	3

For the numerical solution of contact mechanic problems all parameters like penetration and contact stresses are usually not smooth, therefore, a direct distinction of the amount of integration points *a priori* is not possible.

A.1 Gauss Quadrature

A.1.1 Evaluation of Integration Points

Legendre-polynomial of order N:

$$p_N(\xi) = \frac{1}{2^N N!} \frac{d^N(\xi^2-1)^N}{d\xi^N} \rightarrow \text{roots} \stackrel{\wedge}{=} \bar{\xi}_i$$

Weights:

$$w_i = \frac{2}{(1-\bar{\xi}_i^2)(\frac{d\,p_N(\bar{\xi}_i)}{d\xi})^2}$$

Error estimation:

$$R = \frac{2^{(2N+1)}(N!)^4}{(2N+1)[(2N)!]^3} \underbrace{g_{\xi,\xi...}(\bar{\xi})}_{2N}$$

A.1.2 Numerical Examples

1-point integration in $-1 \leq \xi \leq +1$

$\underline{N = 1}$:

$$p_1(\xi) = \frac{1}{2}\frac{d(\xi^2 - 1)}{d\xi} = \xi \implies \text{root at } \bar{\xi}_0 = 0$$

$$w_0 = \frac{2}{(1-0)1^2} = 2$$

$\Rightarrow$ "exact integral"

$$\int_{-1}^{+1} g(\xi)d\xi = 2 \cdot g(0) + R_1$$

$$\text{with} \quad R_1 = \frac{2^3 \cdot 1}{(2 \cdot 1 + 1)2^3}\frac{d^2 g(\hat{\xi})}{d\xi^2} = \frac{1}{3}\frac{d^2 g(\hat{\xi})}{d\xi^2}$$

2-point integration in $-1 \leq \xi \leq +1$

$\underline{N = 2}$:

$$p_2(\xi) = \frac{1}{2^2 \cdot 2 \cdot 1}\frac{d^2(\xi^2-1)^2}{d\xi^2} = \frac{1}{8}\frac{d(2 \cdot 2\xi(\xi^2-1))}{d\xi} = \frac{1}{2}\frac{d}{d\xi}(\xi^3 - \xi) = \frac{1}{2}(3\xi^2 - 1)$$

$$\implies \text{roots at } \bar{\xi}_{0,1} = \pm\frac{1}{\sqrt{3}}$$

$$w_{0,1} = \frac{2}{\left(1-\frac{1}{3}\right)\left[3\left(\pm\frac{1}{\sqrt{3}}\right)\right]^2} = 1 \quad \text{with} \quad \frac{d\,p_2(\xi)}{d\xi} = \frac{d}{d\xi}\left[\frac{1}{2}(3\xi^2 - 1)\right] = 3\xi$$

$\Rightarrow$ "exact integral"

$$\int_{-1}^{+1} g(\xi)d\xi = 1 \cdot g\left(-\frac{1}{\sqrt{3}}\right) + 1 \cdot g\left(+\frac{1}{\sqrt{3}}\right) + R_2$$

$$\text{with} \quad R_2 = \frac{2^5 \cdot 2^4}{5(4 \cdot 3 \cdot 2)^3}\frac{d^4 g(\hat{\xi})}{d\xi^4} = \frac{1}{135}\frac{d^4 g(\hat{\xi})}{d\xi^4}$$

Appendix B

Higher Order Shape Functions of Different Classes

B.1 General

Figure B.1 gives the quadratic shape functions of different classes for the 1D case within the range -1 to 1 as usually used for FEM.

B.2 Lobatto Class

B.2.1 1D Lobatto

Legendre polynomials can be described as

$$L_n(x) = \frac{1}{2^n n!} \frac{d^n}{dx^n} (x^2 - 1)^n, \qquad n = 0, 1, 2, \ldots, \tag{B.1}$$

where n denotes the order of the polynomial and the first four Legendre polynomials are then

$$L_0(x) = 1, \tag{B.2}$$

$$L_1(x) = x, \tag{B.3}$$

$$L_2(x) = \frac{1}{2}(3x^2 - 1), \tag{B.4}$$

$$L_3(x) = \frac{1}{2}x(5x^2 - 3). \tag{B.5}$$

Introduction to Computational Contact Mechanics: A Geometrical Approach, First Edition.
Alexander Konyukhov and Ridvan Izi.
© 2015 John Wiley & Sons, Ltd. Published 2015 by John Wiley & Sons, Ltd.
Companion Website: www.wiley.com/go/Konyukhov

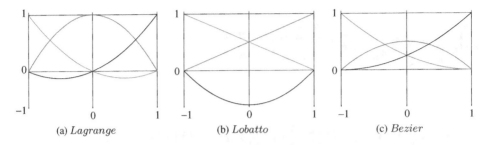

Figure B.1 1D quadratic shape functions

Since they are not appropriate for FEM use the integrated version – Lobatto polynomials – are used instead

$$l_0(x) = \frac{1-x}{2}, \tag{B.6}$$

$$l_1(x) = \frac{1+x}{2}, \tag{B.7}$$

$$l_n(x) = \sqrt{\frac{2n-1}{2}} \int_{-1}^{x} L_{n-1}(\xi)\, d\xi, \quad n = 2, 3, \ldots. \tag{B.8}$$

where for higher order polynomials

$$\int_{-1}^{1} L_k(x)\, dx = 0, \quad k \geq 1 \tag{B.9}$$

holds. The used quadratic Lobatto polynomial can then be expressed as

$$l_2(x) = \frac{1}{2}\sqrt{\frac{3}{2}}(x^2 - 1). \tag{B.10}$$

For clarification, Figure B.2 demonstrates the geometry representation with quadratic shape functions for the Lagrange and Lobatto class. Here, the nodal values $(-1, 0)$, $(0, 1)$ and $(1, 0)$ are used.

B.2.2 2D Lobatto

The linear basis is denoted with the *vertex functions* ($\varphi^{v_1}, \ldots, \varphi^{v_4}$)

$$\varphi^{v_1}(\xi, \eta) = l_0(\xi)\, l_0(\eta), \tag{B.11}$$

$$\varphi^{v_2}(\xi, \eta) = l_1(\xi)\, l_0(\eta), \tag{B.12}$$

$$\varphi^{v_3}(\xi, \eta) = l_1(\xi)\, l_1(\eta), \tag{B.13}$$

$$\varphi^{v_4}(\xi, \eta) = l_0(\xi)\, l_1(\eta). \tag{B.14}$$

Appendix B: Higher Order Shape Functions of Different Classes

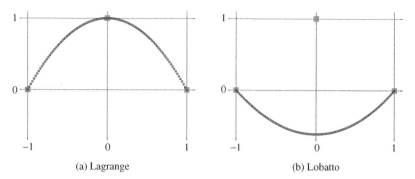

(a) Lagrange   (b) Lobatto

Figure B.2 Line representation with quadratic Lagrange and Lobatto shape functions using (square) nodal values

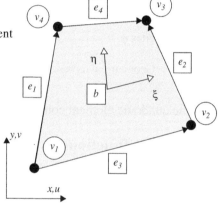

For the 2D case of a plane element, the alignment of shape functions is given according to Solin and Segeth (2004) and is represented here within the reference element.

Figure B.3 High order finite element and alignment of degree of freedoms – vertex, edge and bubble degrees of freedom

where additionally the *edge functions* ($\varphi_n^{e_j}$, $n = 2, \ldots, p^{e_j}$, $j = 1, \ldots, 4$) are used for the higher order.

$$\varphi_k^{e_1}(\xi, \eta) = l_0(\xi)\, l_n(\eta), \quad 2 \leq n \leq p^{e_1}, \tag{B.15}$$

$$\varphi_k^{e_2}(\xi, \eta) = l_1(\xi)\, l_n(\eta), \quad 2 \leq n \leq p^{e_2}, \tag{B.16}$$

$$\varphi_k^{e_3}(\xi, \eta) = l_n(\xi)\, l_0(\eta), \quad 2 \leq n \leq p^{e_3}, \tag{B.17}$$

$$\varphi_k^{e_4}(\xi, \eta) = l_n(\xi)\, l_1(\eta), \quad 2 \leq n \leq p^{e_4}. \tag{B.18}$$

Finally, the *bubble functions* are added as

$$\varphi_{i,j}^{b}(\xi, \eta) = l_i(\xi)\, l_j(\eta), \quad 2 \leq i \leq p^{b_1},\ 2 \leq j \leq p^{b_2}. \tag{B.19}$$

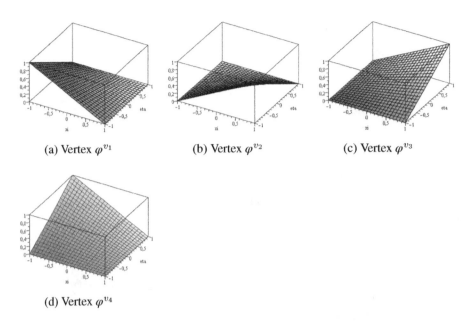

(a) Vertex φ^{v_1} (b) Vertex φ^{v_2} (c) Vertex φ^{v_3}

(d) Vertex φ^{v_4}

Figure B.4 Geometrical representation of the shape in 3D: linear 2D Lobatto shape functions

Thus, the quadratic element consists of the following shape functions:

Linear shape functions

Vertex shape functions:

$$\varphi^{v_1} = \frac{1}{4}(1-\xi)(1-\eta) \qquad \varphi^{v_3} = \frac{1}{4}(1+\xi)(1+\eta)$$

$$\varphi^{v_2} = \frac{1}{4}(1+\xi)(1-\eta) \qquad \varphi^{v_4} = \frac{1}{4}(1-\xi)(1+\eta)$$

Quadratic shape functions

Additional shape functions for quadratic element edge shape functions (see Figure B.5(a)–(d)):

$$\varphi_2^{e_1} = \frac{1}{4}\sqrt{\frac{3}{2}}(1-\xi)(\eta^2-1) \qquad \varphi_2^{e_3} = \frac{1}{4}\sqrt{\frac{3}{2}}(1-\eta)(\xi^2-1)$$

$$\varphi_2^{e_2} = \frac{1}{4}\sqrt{\frac{3}{2}}(1+\xi)(\eta^2-1) \qquad \varphi_2^{e_4} = \frac{1}{4}\sqrt{\frac{3}{2}}(1+\eta)(\xi^2-1)$$

Bubble shape functions (see Figure B.5(e)):

$$\varphi_{2,2}^b = \frac{3}{8}(\xi^2-1)(\eta^2-1)$$

Appendix B: Higher Order Shape Functions of Different Classes

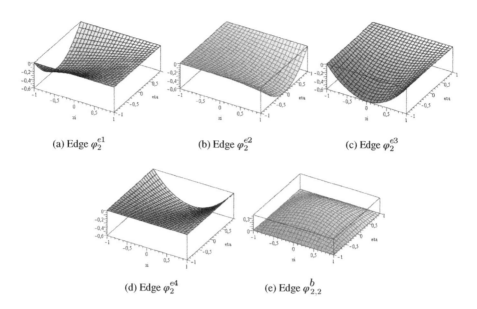

(a) Edge φ_2^{e1} (b) Edge φ_2^{e2} (c) Edge φ_2^{e3}

(d) Edge φ_2^{e4} (e) Edge $\varphi_{2,2}^{b}$

Figure B.5 Quadratic 2D additional Lobatto shape functions

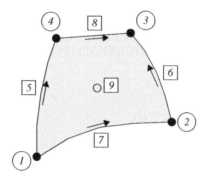

Figure B.6 Used numbering for quadratic Lobatto plane element

B.2.3 Nodal FEM Input

For the FEM analysis the nodal input is given as in Figure B.6 for the quadratic Lobatto plane element.

B.3 Bezier Class

B.3.1 1D Bezier

Bezier polynomials of order n can be expressed as

$$B_i^n(x) = \frac{1}{2^n} \binom{n}{i} (1-x)^{n-i}(1+x)^i, \qquad i = 0, \ldots, n, \qquad \text{(B.20)}$$

where the properties

$$B_i^n(x) = B_{n-i}^n(-x) \tag{B.21}$$

$$\sum_{i=0}^{n} B_i^n(x) = 1 \tag{B.22}$$

hold.

Thus, the Bezier shape functions are for the linear case:

$$B_0^1(x) = \frac{1}{2}(1-x), \tag{B.23}$$

$$B_1^1(x) = \frac{1}{2}(1+x), \tag{B.24}$$

quadratic case:

$$B_0^2(x) = \frac{1}{4}(1-x)^2, \tag{B.25}$$

$$B_1^2(x) = \frac{1}{2}(1-x)(1+x), \tag{B.26}$$

$$B_2^2(x) = \frac{1}{4}(1+x)^2. \tag{B.27}$$

As for the Lobatto polynomial the geometry representation for the Bezier class is clarified using nodal points as $(-1, 0)$, $(0,1)$ und $(1,0)$ see Figure B.7.

B.3.2 2D Bezier

The *vertex functions* $\varphi^{v_1}, \ldots, \varphi^{v_4}$ are

$$\varphi^{v_1}(\xi, \eta) = B_0^n(\xi) \, B_0^n(\eta), \tag{B.28}$$

$$\varphi^{v_2}(\xi, \eta) = B_n^n(\xi) \, B_0^n(\eta), \tag{B.29}$$

$$\varphi^{v_3}(\xi, \eta) = B_n^n(\xi) \, B_n^n(\eta), \tag{B.30}$$

$$\varphi^{v_4}(\xi, \eta) = B_0^n(\xi) \, B_n^n(\eta), \tag{B.31}$$

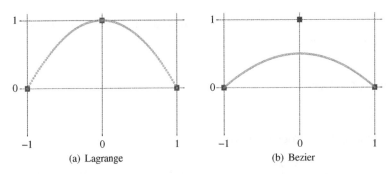

Figure B.7 Line representation with quadratic Lagrange and Bezier shape functions using (square) nodal values

Appendix B: Higher Order Shape Functions of Different Classes

while the *edge functions* $\varphi_i^{e_j}$ can be expressed as

$$\varphi_i^{e_1}(\xi,\eta) = B_0^n(\xi)\, B_i^n(\eta), \quad 1 \leq k \leq n-1, \tag{B.32}$$

$$\varphi_i^{e_2}(\xi,\eta) = B_n^n(\xi)\, B_i^n(\eta), \quad 1 \leq k \leq n-1, \tag{B.33}$$

$$\varphi_i^{e_3}(\xi,\eta) = B_i^n(\xi)\, B_0^n(\eta), \quad 1 \leq k \leq n-1, \tag{B.34}$$

$$\varphi_i^{e_4}(\xi,\eta) = B_i^n(\xi)\, B_n^n(\eta), \quad 1 \leq k \leq n-1, \tag{B.35}$$

and, the *bubble functions*

$$\varphi_{i,j}^b(\xi,\eta) = B_i^n(\xi) B_j^n(\eta), \quad 1 \leq i \leq n-1, \quad 1 \leq j \leq n-1. \tag{B.36}$$

The linear case for the Bezier class of shape functions is the same as for the Lagrange and Lobatto class, therefore, only the quadractic shape functions are given further.

Quadratic shape functions

Vertex shape functions (e.g. see Figure B.9(a)):

$$\varphi^{v_1} = \frac{1}{16}(1-\xi)^2(1-\eta)^2 \qquad \varphi^{v_3} = \frac{1}{16}(1+\xi)^2(1+\eta)^2$$

$$\varphi^{v_2} = \frac{1}{16}(1+\xi)^2(1-\eta)^2 \qquad \varphi^{v_4} = \frac{1}{16}(1-\xi)^2(1+\eta)^2$$

Edge shape functions (e.g. see Figure B.9(b)):

$$\varphi_1^{e_1} = \frac{1}{8}(1-\xi)^2(1-\eta^2) \qquad \varphi_1^{e_3} = \frac{1}{8}(1-\eta)^2(1-\xi^2)$$

$$\varphi_1^{e_2} = \frac{1}{8}(1+\xi)^2(1-\eta^2) \qquad \varphi_1^{e_4} = \frac{1}{8}(1+\eta)^2(1-\xi^2)$$

For the 2D case of a plane element the alignment of shape functions is represented here within the reference element.

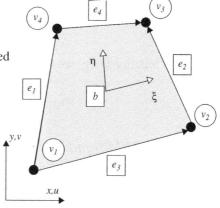

Figure B.8 Bi-quadratic plane finite element. Alignment of degree of freedoms – nodal degrees of freedom

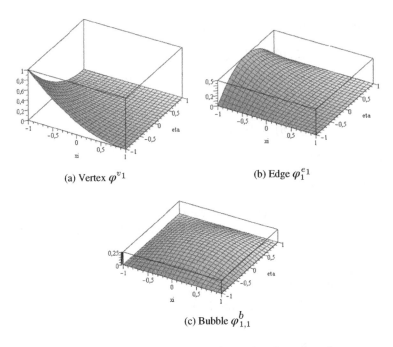

Figure B.9 Selection of quadratic Bezier shape functions

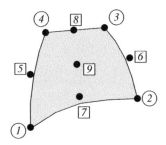

Figure B.10 Used numbering for quadratic Bezier plane element

Bubble shape functions (e.g. see Figure B.9(c)):

$$\varphi^b_{1,1} = \frac{1}{4}(1-\xi^2)(1-\eta^2)$$

B.3.3 Nodal FEM Input

For the FEM analysis the nodal input is given as in Figure B.10 for the quadratic Bezier plane element.

References

Bonet, J., Wood, R. D. 1999. *Non-Linear Continuum Mechanics for Finite Element Analysis.* Cambridge University Press.

Hughes Th. J. R. *The Finite Element Method: Linear Static and Dynamic Finite Element Analysis.* Dover Publications.

Johnson K. L. 1987. *Contact Mechanics.* Cambridge University Press.

Kikuchi N., Oden J. T. 1988. *Contact Problems in Elasticity: A Study of Variational Inequalities and Finite Element Methods.* SIAM.

Konyukhov A. 2013. Contact of ropes and orthotropic rough surfaces, ZAMM, *Journal of Applied Mathematics and Mechanics*, DOI: 10.1002/zamm.201300129.

Konyukhov A., Schweizerhof K. 2012. *Computational Contact Mechanics Geometrically Exact Theory for Arbitrary Shaped Bodies.* Springer.

Laursen T. 2002. *Computational Contact and Impact Mechanics Fundamentals of Modeling Interfacial Phenomena in Nonlinear Finite Element Analysis.* Springer.

Popov V. L. 2010. *Contact Mechanics and Friction: Physical Principles and Applications.* Springer.

Sofonea M., Matei A. 2012. *Mathematical Models in Contact Mechanics.* Cambridge University Press.

Solin P., Segeth K., Dolezel I. 2004. *Higher Order Finite Element Methods.* Chapman and Hall.

Wriggers P. 2002. *Computational Contact Mechanics.* John Wiley & Sons, Ltd.

Yastrebov A. 2013. *Numerical Methods in Contact Mechanics.* Wiley-ISTE.

Index

absolute value, 35
abstract form of formulations in computational mechanics, 61–71
   contact formulations, 69–71, *See also individual entry*
   fixed point theorem (Banach), 64–65
   functional –strain energy operator for, 62
   functional operator for, 62
   iterative method, 63
   linear matrix operator for, 61
   Newton iterative solution method, 65–68
   operator necessary for, 61–62
   rate of convergence, 64
allowable projection domain, 49
   for circular arch, 50–52
analytical solutions, verification with, 109–120, *See also* Hertz problem; rigid flat punch problem
   generalized Euler–Eytelwein problem, 118–120
   impact on moving pendulum, center of percussion, 116–118
angle between two curves, 31
arbitrary two body contact problem, 45–59

contact kinematics, 55–59
   geometry for, 45–59
   kinematics for, 45–59
   local coordinate system, 46–48
area of surface, 32
augmented Lagrangian method, 8–11

Babuska–Brezzi condition, 70
basis vectors, 30–35
Bezier class, 269–272
   1D Bezier, 269–270
   2D Bezier, 270–272
bubble functions, 271

C1-continuity, 52
Cauchy stress theorem, 16
center of percussion, 116–118
Christoffel symbols, 38, 42–43
circle and its properties, 26–28
circular arch, allowable projection domain for, 50–52
closed form solution for STAS penetration, 99
   contact with a rigid line, 99
   in 2D, 100–101
   master surface as an analytical rigid straight line, 100–101

Introduction to Computational Contact Mechanics: A Geometrical Approach, First Edition.
Alexander Konyukhov and Ridvan Izi.
© 2015 John Wiley & Sons, Ltd. Published 2015 by John Wiley & Sons, Ltd.
Companion Website: www.wiley.com/go/Konyukhov

closed form solution for STAS penetration (*continued*)
   master surface as analytical rigid circle, 101
Closest Point Projection (CPP), 45, 48–55
   allowable projection domain for the circular arch, 50–52
   analysis, 48–55
   in 3D, 54–55
   existence of, 49–50
   irregular cases, 52–55
   numerical solution in 2D, 54
   uniqueness of, 49–50
computational mechanics, 61–71, *See also* abstract form of formulations
consistent linearization, 75–79
   linearization of normal part, 76–79
constitutive equations, 15
constitutive laws of material, 148–149, 158
contact formulations, 5–11
   Lagrange multiplier method, 5–6
   penalty method, 6–11
   augmented Lagrangian method, 8–11
contact formulations, abstract form for, 69–71
   Babuska–Brezzi condition, 70
   Karush–Kuhn–Tucker conditions, 69
   Lagrange multiplier method in operator form, 69
   penalty method in operator form, 71
   Signorini problem, 70
contact geometry, 110–113
contact integral for contact approaches, computation of, 86–87
contact interactions, measures of, 121–123
contact kinematics, 55–59
   for 2D case, 56
   in 3D case, 57
   in 3D coordinate system, 59
   using natural coordinates, 58

contact patch test, 209
contact problem, 13–21
   contact part formulation (Signorini's problem), 17–21
   general formulation of, 13–21
   strong formulation of equilibrium, 14
   structural part, formulation of a problem in linear elasticity, 13–17
contravariant basis vectors, 33–34
convective variation, linearization of, 81
convergence of Newton method, 67
convex function, 67
Coulomb friction law, 123
covariant basis vectors, 33–34
covariant components of metric tensor, 41
covariant derivatives
   on cylindrical surface, 43–44
   on surface, 38–39
covariant tangent vectors, 30
curvature of a curve, 25
curvature tensor, 35–37
   analysis of, generalized Eigenvalue problem, 36–37
   components, 41–42
   Gaussian curvature, 36
   mean curvature, 36
   and surface structure, 35–37
curve and its properties, 23–28
   circle, 26–28
   Frenet formulas in 2D, 28

1D Bezier, 269–270
1D Lobatto, 265–266
2D
   closed form solutions for STAS penetration in, 100–101
   contact kinematics for 2D case, 56
   CPP procedure in, numerical solution of, 54
   Frenet formulas in, 28
2D Bezier, 270–272

bubble functions, 271
edge functions, 271
nodal FEM input, 272
quadratic shape functions,
   271–272
vertex functions, 270
2D coordinate system, 47
2D Hertz solution, 114
   for cylinders, 114
2D Lobatto, 266–269
3D, 54–55
   contact kinematics in, 57
   CPP procedure in, 54–55
3D coordinate system, contact
   kinematics in, 59
3D Hertz solution, 113–114
   for spheres, 113–114
3D Node-To-Segment (NTS)
   (elmt107.f), 221–227
   bending over a rigid cylinder, 227
   bending over a rigid sphere, 227
   description of subroutines, 223–225
   geometry of, 223
   global FEAP-arrays, 225
   implementation, 222–225
   parameters of, 223
   sliding on a ramp, 226
   two blocks, 3D case, 226
differential geometry, 23–44
   curve and its properties, 23–28
differential properties of surfaces,
   37–44
   covariant components of metric
      tensor, 41
   covariant derivatives on a cylindrical
      surface, 43–44
   covariant derivatives on the surface,
      38–39
   curvature tensor components, 41–42
   Gauss–Codazzi formula, 38
   Weingarten formula, 37–38
discretization of frictional NTN,
   131–133

edge functions, 271
Einstein notation, 14
Einstein summation convention, 14
elasto-plastic analogy, 125
equilibrium equation, 14

FEAP (Finite Element Analysis
   Program)
finite element analysis program, 141
finite element discretization, 85–108,
      See also Node-To-Segment (NTS)
      contact element
   contact integral for, computation of,
      86–87
finite element method (FEM), 66
first fundamental form of a surface, 30
fixed point theorem (Banach), 64–65
Frenet formulas, 79
Frenet frame, 24
   first Frenet formula, 24
friction law, 121–123
   Coulomb friction law, 123
frictional contact patch test, 233–234,
   241–242
frictional contact problems, 121–136
   measures of contact interactions,
      friction law, 121–123
   weak form and its consistent
      linearization, 128–129
frictional higher order NTS (elmt110.f),
   245–253
   block and inclined rigid surface,
      252–253
   description of subroutines, 248–250
   geometry of, 245
   global FEAP-arrays, 250
   implementation, 246–250
   parameters of, 245
   two blocks, 251–252
frictional Node-To-Node (NTN)
   (elmt108.f), 229–234
   description of subroutines, 231
   frictional contact patch test, 233–234

frictional Node-To-Node (NTN)
(elmt108.f) (*continued*)
 geometry of, 231
 global FEAP-arrays, 232
 implementation, 230–232
 parameters of, 231
 two blocks, 232–233
frictional Node-To-Node (NTN) contact element, 129–134
 discretization of, 131–133
 full residual for, 132
 linearization of tangential part for, 131
 local level, algorithm for, 133–134
 regularization of contact conditions, 130–131
frictional Node-To-Segment (NTS) (elmt109.f), 235–244
 block and inclined rigid surface, 242
 description of subroutines, 238–239
 frictional contact patch test, 241–242
 generalized 2D Euler–Eytelwein problem, 243–244
 geometry of, 238
 global FEAP-arrays, 239
 implementation, 236–240
 parameters of, 238
 two blocks, 240–241
frictional Node-To-Segment (NTS) contact element, 134–135
 discretization for, 134–135
 linearization for, 134–135
 local level, algorithm for, 135
 NTS frictional contact element, 135–136
full residual for frictional NTN contact element, 132

Gauss–Codazzi formula, 38
Gauss coordinates, surfaces description by, 29–37
 basis vectors, metric tensor and its applications, 30–35
 curvature tensor and structure of surface, 35–37
 surface coordinate system, tangent and normal vectors, 29–30
Gauss point-wise substituted formulation, 82
Gauss quadrature, 262–263
 integration points evaluation, 262
 numerical examples, 263
Gauss theorem, 16
Gaussian curvature, 36
generalized 2D Euler–Eytelwein problem, 243–244
generalized Eigenvalue problem, 36–37
generalized Euler–Eytelwein problem, 118–120
 rope on a circle, 119–120
 rope on an ellipse, 119–120
geometrical interpretation of Newton iterative method, 66–68

heaviside function, 82
Hertz problem, 109–114, 183–184
 contact condition for, 112
 contact geometry, 110–113
 contacting surfaces for, 111
 2D Hertz solution, 114
 3D Hertz solution, 113–114
 non-penetration condition for, 112
higher order Mortar/STS (elmt106.f), 213–220
 block and inclined rigid surface, 219–220
 description of subroutines, 215–217
 geometry of, 215
 global FEAP-arrays, 217
 implementation, 214–217
 parameters of, 215
 two blocks, 218–219
higher order shape functions of different classes, 265–272

Bezier class, 269–272
general, 265
Lobatto class, 265–269

integration points evaluation, 262
inverted contact algorithm, 103, 140, 177, 185–187, 196–201, 203, 210–212
　description of subroutines, 197
　distributed following forces, pressure, 199–201
　following force, 185–187
　implementation, 197, 210–212
　inflating of a bar, 201
　verification of a rotational part, 198–199, 211–212
irregular cases, 52–55
iterative method, abstract form of, 63

Karush–Kuhn–Tucker (KKT) conditions, 6, 13, 69

Lagrange multiplier method, 5–6
Lagrange multiplier Node-To-Node (NTN) (elmt101.f), 165–169
　description of subroutines, 167
　geometry of, 167
　global FEAP-arrays, 168
　implementation, 166–168
　parameters of, 167
　three trusses, 169
　two trusses, 168
Lagrange multipliers, 79–81
　linearization for, 80
　penalty method application to, 79–81
linear elastic material, 15
linear elasticity, formulation of a problem in, 13–17
　strong formulation of equilibrium, 14
　weak formulation of equilibrium, 15–17
linear infinitesimal stain tensor, 15
linear shape functions, 268

linearization of convective variation, 81
linearization of normal part, 76–79
linearization of tangential part for the NTN contact approach, 131
Lobatto class, 265–269
　1D Lobatto, 265–266
　2D Lobatto, 266–269
　linear shape functions, 268
　nodal FEM input, 269
　quadratic shape functions, 268–269
Lobatto type integration, 96
local coordinate system, 46–48
　weak formulation in, 73–75
local level frictional NTN contact element, algorithm for, 133–134
local level NTS frictional contact element, algorithm for, 135

master surface
　as analytical rigid circle, 101
　as an analytical rigid straight line, 100–101
mean curvature, 36
metric tensor and its applications, 30–35
　angle between two curves, 31
　area of surface, 32
　contravariant basis vectors, 33–34
　covariant basis vectors, 33–34
　length of a curve laying on a surface, 31
　metric properties of surface, 31
metric tensor, covariant components of, 41
Mortar/Segment-To-Segment (STS) (elmt105.f), 203–212
　block and inclined rigid surface, 208–209
　contact patch test, 209
　description of subroutines, 205–206
　geometry of, 205
　global FEAP-arrays, 206
　implementation, 204–207

Mortar/Segment-To-Segment (STS)
   (elmt105.f) (*continued*)
  inverted contact algorithm, 210–212
  parameters of, 205
  two blocks, 207–208

Nabla operator, 14
Newton iterative solution method,
   65–68
  convergence of, 67
  geometrical interpretation of, 66–68
Nitsche method, 81–83
Nitsche Node-To-Node (NTN)
   (elmt102.f), 171–175
  description of subroutines, 173
  geometry of, 173
  global FEAP-arrays, 174
  implementation, 171–174
  parameters of, 173
  three trusses, 174–175
  two trusses, 174
Nitsche Node-To-Node (NTN) contact
   element, 89–91
Node-To-Node (NTN) approach, 86
  numerical integration for, 86
Node-To-Node (NTN) contact element,
   88–89
Node-To-Segment (NTS) (elmt103.f),
   177–187
  description of subroutines, 179
  geometry of, 179
  global FEAP-arrays, 180
  Hertz problem, 183–184
  implementation, 178–181
  inverted contact algorithm, following
     force, 185–187
  parameters of, 179
  two blocks, 181–182
  two cantilever beams, large sliding
     test, 183
Node-To-Segment (NTS) approach, 86
  master and slave parts for, selection
     of, 96

  numerical integration for, 86
Node-To-Segment (NTS) contact
   element, 91–98
  computation of, peculiarities in,
     95–96
  geometry of, 93
  kinematics of, 93
  linear NTS contact element, closest
     point projection procedure for,
     94–95
  main part necessity, 97
  residual matrix, 96–98
  tangent matrix, 96–98
nonlinear structural plane (elmt2.f),
   151–158
  constitutive law of material, 158
  description of subroutines, 154–155
  geometry of, 151
  global FEAP-arrays, 155
  implementation, 152–156
  large rotation, 158
  parameters of, 151
nonlinear structural truss element
   (elmt1.f), 143–150
  constitutive laws of material,
     148–149
  description of subroutines, 146
  geometry of, 145
  global FEAP-arrays, 147
  implementation, 144–147
  large rotation, 149
  parameters of, 145
  snap-through buckling, 150
non-penetration condition, 19
normal vector, 27
normalized initial gap, 19
numerical integration, 261–263

penalty method, 6–11, 75
  application to Lagrange multipliers,
     79–81
  mechanical interpretation of, 7
  regularization with, 75

penalty method in operator form, 71
penalty Node-To-Node (NTN)
    (elmt100.f), 159–163
  description of subroutines, 161
  geometry of, 160
  global FEAP-arrays, 161
  implementation, 160–161
  parameters of, 160
  three trusses, 162
  two blocks, 162–163
  two trusses, 161
principle of maximum of dissipation, 125
programming and verification tasks, 139–142

quadratic shape functions, 268–269, 271–272

rate of convergence, 64
residual matrix, 96–98, 106–108
residual vector, 66
return mapping algorithm, 123–128
  elasto-plastic analogy, principle of maximum of dissipation, 125
  regularization of, 123–128
  sliding displacements update in reversible loading, 127–128
rigid flat punch problem, 114–115
rotational part, 97

saddle point, 69
Segment-To-Analytical-Segment (STAS) (elmt104.f), 189–201
  bending over a rigid cylinder, 196
  block and inclined rigid surface, 194–195
  block and rigid surface, 193–194
  description of subroutines, 191–192
  global FEAP-arrays, 192
  implementation, 190–193

Segment-To-Analytical-Segment (STAS) approach, 86, 98–103
  closed form solution for, 99
  CPP procedure for STAS contact element, general structure of, 98–100
  discretization for, 102
  geometry of, 99
  kinematics of, 99
  numerical integration for, 86–87
  residual matrix, 102–103
  tangent matrix, 102–103
Segment-To-Segment (STS) Mortar approach, 86, 104–108
  CPP procedure for, peculiarities of, 106
  numerical integration for, 87
  residual matrix, computation of, 106–108
  tangent matrix, computation of, 106–108
Signorini's problem, 17–21, 70
  contact with rigid obstacle, 18
  non-penetration condition, 19
  normalized initial gap, 19
sliding displacements update in reversible loading, 127–128
snap-through buckling, 150
spring-mass frictionless contact system, 3–11, *See also* contact formulations
  contact part, non-penetration into rigid plane, 4–5
  with a rigid plane, 4
  structural part, spring-mass system deflection, 3–4
stabilization parameter, 83
surface area, 32
surface coordinate system, tangent and normal vectors, 29–30
surfaces, 37–44, *See also* differential properties of surfaces

tangent matrix, 66, 96–98, 106–108
tangential force, regularization, 123–128
transient contact problems, 255–259
   block and inclined rigid surface, 257, 258
   description of subroutines, 257
   implementation, 256–257

truss element, 143–150, *See also* nonlinear structural truss element

vertex functions, 270

weak formulation, 73–83
   in local coordinate system, 73–75
Weingarten formula, 37–38, 59

Printed and bound by CPI Group (UK) Ltd, Croydon, CR0 4YY
09/06/2025
14685658-0001